Contents

Drink, Drugs and Dependence

From science to clinical practice

Edited by Woody Caan and
Jackie de Belleroche

London and New York

First published 2002
by Routledge
11 New Fetter Lane, London EC4P 4EE

Simultaneously published in the USA and Canada
by Routledge
29 West 35th Street, New York, NY 10001

Routledge is an imprint of the Taylor & Francis Group

© 2002 Routledge 2002

Typeset in Times by Steven Gardiner Ltd
Printed and bound in Great Britain by MPG Books Ltd, Bodmin

British Library Cataloguing in Publication Data
A catalogue record for this book is available
from the British Library

Library of Congress Cataloging in Publication Data
A catalog record has been requested

ISBN 0415-27891-0 (hbk)
      0415-27901-1 (pbk)

Figures and illustrations

Tables

r

Contributors

Heather Ashton is Professor of Psychiatry at the Department of Psychiatry, University of Newcastle upon Tyne, Leazes Wing, Royal Victoria Infirmary, Newcastle upon Tyne NE1 4LP.

Alexander Baldacchino is Senior Lecturer in Psychiatry of Addictive Behaviour at the Department of Psychiatry, University of Dundee, Ninewells Hospital and Medical School, Dundee DD1 9SY.

Woody Caan is Public Health Specialist in Research and Development at the International Centre for Health and Society, University College London, London WC1E 6BT.

John Chacksfield is Head of Therapy and Education, John Howard Centre, Therapy and Education Department, 2 Crozier Terrace, Hackney, London E9 6AT.

Jackie de Belleroche is Professor of Neurochemistry at the Division of Neurosciences and Psychological Medicine, Faculty of Medicine, Imperial College School at Charing Cross Hospital, Fulham Palace Road, London W6 8RF.

Graham Divall is Deputy General Manager (Drugs) at the Forensic Science Service London Laboratory, 109 Lambeth Road, London SE1 7LP.

Sue Drummond is Teaching Fellow at the Centre for Research on Drugs and Health, Department of Social Science and Medicine, Imperial College School of Medicine, Reynolds Building, Charing Cross Hospital, St Dunstan's Road, London W6 8RP.

Hilary J. Little is Professor of Psychology at the Drug Dependence Unit, Psychology Department, University of Durham, South Road, Durham DH1 3LE.

E. Jane Marshall is Senior Lecturer at the National Addiction Centre, 4 Windsor Walk, London SE5 8AF.

John McCartney is Senior Lecturer in Psychology at the Psychology Department, Calcutta House, Old Castle Street, London Guildhall University, London E1 7NT and Honorary Clinical Psychologist at St George's Hospital, Cranmer Terrace, London SW17 0RE.

Timothy J. Peters is Professor in Clinical Biochemistry at the Department of Clinical Biochemistry, King's College School of Medicine and Dentistry, London SE5 9PJ.

Janet T. Powell is Professor of Vascular Biology at the Departments of Biochemistry and Surgery, Charing Cross and Westminster Medical School, St Dunstan's Road, London W6 8RP.

Victor R. Preedy is Reader in Clinical Biochemistry at the Department of Clinical Biochemistry, King's College School of Medicine and Dentistry, London SE5 9PJ.

Marcus Rattray is Senior Lecturer in Biochemistry at the Biochemical Neuropharmacology Group, Centre for Neuroscience Research, GKT School of Biomedical Sciences, King's College London, Hodgkin Building, Guy's Hospital Campus, London SE1 1UL.

John Shanks is Director of Public Health at Croydon Health Authority, Knollys House, Croydon, Surrey, CR0 6SR.

Teresa D. Tetley, Senior Lecturer in Respiratory Medicine at the National Heart and Lung Institute at Charing Cross Hospital, Imperial College School of Medicine, London W6 8RF.

Nicola Griffiths is a member of the Tobacco Policy Team, Department of Health, Room 432, Wellington House, London SE1 8UG.

William P. Watson is Section Leader of Neuropharmacology (805), H. Lundbeck A/S, Olliliavej 9, DK-2500 Valby, Denmark.

Foreword

I was delighted to accept the invitation to provide a foreword for this publication which is targeted at professionals first encountering individuals with drug and alcohol related problems.

The government's anti-drug strategy, set out in *Tackling Drugs to Build a Better Britain*, seeks to shift the emphasis away from dealing with the consequences of drug misuse to preventing them happening in the first place. Education has a key role to play, as does ensuring that those who develop drug problems can access good, effective treatment programmes. To help realize our goals, significant additional resources have been made available but equally important government departments, agencies and others in the field are working closely together towards common targets.

The editors of this publication have also recognized the benefits of effective co-operation, and the range of disciplines that have contributed to this publication is impressive. It combines the scientific with the practical and presents issues clearly in context and will, I am sure, prove a valuable resource to many.

Only by working effectively together can we achieve our aim of reducing the harm and damage that drugs and alcohol cause to individuals and society.

Keith Hellawell
The UK Anti-drugs Co-ordinator

Preface

The misuse of alcohol and drugs is a major inescapable feature of our society. However, the majority of cases or types of problem go untreated and it is only recently that substance abuse has been recognized as a serious health problem. A measure of the extent of the problem is indicated by the estimated 4 per cent of consumer spending on alcohol and drugs. In the UK, this is approximately £12 billion on alcohol and £10 billion (ten billion pounds sterling) on illegal drugs. Drug use is an international issue affecting all age groups. Increasingly, people are being damaged by substance misuse and every caring profession has to address the problem of drug and alcohol use amongst clients, even if these clients are schoolchildren or pensioners.

Enormous social changes are evolving as seen in the decreased average age of first experimentation in substance misuse from late teens a couple of decades ago to pre- or early teens (11–15 years) in the late 1990s. The range of substances widely available has also expanded greatly over this period from cannabis, amphetamine and LSD to crack cocaine, solvents, ecstasy and its variants. Hand in hand with the expansion in illegal drug availability has come the development of ruthless marketing strategies which introduce new drugs initially very cheaply to ensnare clients.

A number of clinical, scientific and sociological skills need to be drawn on to address the problems of substance misuse. No one discipline has all the answers to understanding or managing this expanding problem. In fact, it is essential to build up a complete picture of substance misuse from the nature of the chemicals being used to their effects on the individual, the family and society as a whole.

The aim of this book is to guide the reader through the nature of drink and drug problems, especially dependence, using a broad perspective: from the basic brain mechanisms to the physical, psychological, sociological and practical issues. We aim to provide an understanding of the whole system in order to prepare anyone in the caring professions for their encounter with chemical dependence and also to provide the basis for responding in the future to inevitable changes in drug taking that will occur. The book starts with an introduction to broad issues relating to population health and policy. We then describe familiar legal substances such as alcohol and tobacco covering the genetic, cellular and psychological effects on the individual and the wider community. We then move on to cover illegal drugs, the problems encountered after short-term use and the effects of dependence. The different approaches to treatment are developed to provide a rationale for their use rather than a recipe. Finally, we scan the horizon for future developments to prepare the reader for rapid changes that occur without warning. These may be novel chemicals or could be old drugs in an unfamiliar setting such as the recent emergence of injectable steroids being encountered in the 'clubbing scene' or the marketing of alcoholic beverages for a young age group. We have tried to harmonize the chapters of the book into a similar format with key point summaries. However, each chapter can be read on its own, it is up to the reader to customize their own requirements.

Abbreviations

AA	Alcoholics Anonymous
ACMD	Advisory Council on the Misuse of Drugs
ADH	Alcohol dehydrogenase
AHMD	Alcoholic Heart Muscle Disease
ALDH2	A mitochondrial isoenzyme of aldehyde dehydrogenase
AMPA	α-amino-3-hydroxyl-5-methyl-4-isoxazole proprionate
ANSA	Association of Nurses in Substance Abuse Services
ARVAC	Association for Research in the Voluntary and Community Sector
AUDIT	An alcohol screening test
BAC	Blood Alcohol Concentration
BMAST	An alcohol screening test
BPDE	Benzo(a)pyrene Diol
CAGE	Cut, Annoyed, Guilty, Eye-opener
CAOT	Canadian Model of Occupational Performance
CART	Cocaine- and Amphctamine-Related Transcript
CBT	Cognitive Behavioural Therapy
COPD	Chronic Obstructive Pulmonary Disease
CR	Conditioned Response
CYP	Cytochrome P450
DARE	Drug Abuse Resistance Education
DAT	Drug Action Team
DIVC	Disseminated Intravascular Coagulation
DRIE	Drinking-Related Locus of Control Scale
DSH	Deliberate Self-harm
DSM	Diagnostic and Statistical Manual
DZ	Dizygotic
EAP	Employee Assistance Programme
EDM	Ego-defence Mechanisms
FAE	Foetal Alcohol Effects
FAS	Foetal Alcohol Syndrome
GABA	Gamma-aminobutyric acid

GC–MS	Gas Chromatography–Mass Spectrometry
GBH	Grievous Bodily Harm
GHB	γ-Hydroxybutyrate
GP	General medical practitioner, in primary care
GST	Glutathione-S-transferase
HALT	Hungry, Angry, Lonely or Tired
HDA	Health Development Agency
HDL2	High Density Lipoprotein type 2
5-HT	5-Hydroxytryptamine
IBC	A health publisher and conference organizer
ICD	International Classification of Diseases
IL-8	Interleukin-8
ISDD	Institute for the Study of Drug Dependence (now called DrugScope)
LSD	Lysergic Acid Diethylamide
MAP	Mandsley Addiction Profile
MBDB	Methylbenzodioxolybutanamine
MDA	Methylenedioxyamphetamine
MDEA	Methylenedioxyethylamphetamine
MDMA	Methylenedioxymethamphetamine
MPTP	1-Methyl-4-phenyl-1,2,3,6-tetrahydropyridine
MRC	Medical Research Council
MRI	Magnetic Resonance Imaging
MUDA	The Misuse of Drugs Act
MZ	Monozygotic
NA	Narcotics Anonymous
NFER	National Foundation for Educational Research
NIAAA	National Institute on Alcoholism and Alcohol Abuse
NMDA	N-Methyl-D-aspartate
NNK	N-Nitrosamino Ketone
NNN	Nitrosonornicotine
NRT	Nicotine Replacement Therapy
NTORS	National Treatment Outcome Research Study
ONS	Office for National Statistics
OPCS	Office of Population Censuses and Surveys (now called ONS)
PAH	Polycyclic Aryl Hydrocarbon
PAT	Paddington Alcohol Test
PET	Positron emission tomography
PMA	Paramethoxyamphetamine
PMN	Polymorphonuclear Neutrophil
RAPS	An alcohol screening test
RPM	Relapse Prevention Matrix
SANE	Substance Abuse a National Emergency (US)
SCODA	Standing Conference on Drug Abuse (now called DrugScope)

SERT	Serotonin Transporter
SLPI	Secretory Leukoprotease Inhibitor
SSRI	Selective Serotonin Re-uptake Inhibitor
TACADE	A drugs charity focusing on support for teachers and schools
THC	Tetrahydocannabinol
TWEAK	An alcohol screening test
UMDS	United Medical and Dental Schools
WHO	World Health Organization

Chapter 1

What is a drug and what is addiction?

John Shanks

Key points

- *For both alcohol and illicit drugs, the patterns of use and the patterns of harm can be related across a defined population.*
- *Careful use of terminology helps to clarify this relationship.*
- *Good-quality epidemiological information about the patterns of drug use and the resultant drug problems can contribute significantly to designing better health policies.*

Introduction

Many societies have a chosen drug or chemical whose use is incorporated in social rituals, such as coming of age, marriage and other rites of passage. In Europe and North America, alcohol fulfils some of these functions for most members of the population. In other cultures, different drugs, such as opium or cannabis, are the socially sanctioned choice.

The use and misuse of alcohol and drugs affects everyone in our society, even those people who do not themselves consume these substances. The alcohol industry is a huge business that provides employment for those involved in agriculture, manufacturing, distribution and serving of alcoholic beverages. In the UK, we spent over £25 billion buying alcoholic drinks in 1994 and about £180 million advertising them. Revenue from taxes on alcohol raised more than £9 billion for the Exchequer. During 1993–94, the UK government spent at least £526 million on tackling drug misuse: about £350 million of this went on law enforcement while about £165 million was spent on education, prevention and treatment. The economic costs of illicit drug consumption have risen rapidly, with over 2 per cent of total consumer spending in the UK (£10 billion: Butler 1998) expended on drugs. The health problems and crime associated with the misuse of alcohol and drugs impose a heavy burden on individuals, families, communities, law enforcement agencies, health and social services; for example, in France the recent report of Pierre Kopp for the Office of Drugs found 2.68 per cent of the gross national product (over £21 billion annually) was spent on social problems caused by alcohol and tobacco (summarized in Webster 1999).

Terminology

Discussion or reading about alcohol and drug use is complicated by the use of terms such as addiction, dependency, abuse and misuse which are often left undefined and which can be employed with differing and overlapping shades of meaning.

Terms in common use:

- misuse
- abuse
- dependency
- addiction
- problem use
- alcoholism

Among drug users themselves, there is a highly local and frequently changing street language to describe specific drugs and methods of taking them. Unfortunately, there is no universally accepted set of terms or definitions but the terminology proposed by the World Health Organization has achieved fairly wide acceptance in official circles.

What is a drug?

A simple definition of a drug would be any substance that alters the functioning of the body or the mind. This definition of a drug would include alcohol and also substances such as tobacco or caffeine. For the purposes of this chapter, we will exclude tobacco and caffeine from further consideration because of their different status in our society. For similar reasons, we will consider alcohol separately from other drugs. The generic term 'substance' is often used to include both alcohol and drugs.

What is addiction?

The term 'addiction' is often used interchangeably with related terms such as 'dependence', 'abuse', 'misuse' or 'problem use'. All these terms can be applied equally to drugs or to alcohol, which has, additionally, its own specific term 'alcoholism'. The World Health Organization in its disease classification schemes has proposed some distinct categories, which may be useful.

Drug abuse

Persistent or sporadic excessive use of a drug inconsistent with, or unrelated to, medical practice.

Drug dependence

A psychic and sometimes physical state, resulting from taking a drug, characterized by behavioural and other responses that always include a compulsion to take the drug on a continuous or periodic basis in order to experience its psychic effects and sometimes to avoid the discomfort of its absence. This rather unwieldy definition of drug dependence introduces the notion of withdrawal effects resulting from withholding the accustomed drug and implies that the state of drug dependence can be either psychological or physical or both.

Definitions of terms such as these may not be adequate to decide in any particular instance whether an individual can be regarded as having a significant problem with drug or alcohol use. There are, therefore, a range of operational definitions which allow people to be more precisely categorized on the basis of some quantified measure of drug or alcohol use which is then compared with a threshold level agreed to constitute problem use. Population surveys of alcohol consumption, for example, may count the numbers of people who report drinking at levels above the current upper limits of sensible consumption recommended by the UK's medical royal colleges (21 units per week for men, 14 units per week for women). There are also a variety of standardized questionnaires, which produce a score based on the presence of identifiable characteristics thought to be associated with definite alcohol or drug misuse. The CAGE questionnaire is one such short questionnaire, which enquires about four characteristics that are judged to occur in moderately severe problem drinking. A positive response to two or more of the four questions is often taken to indicate a likely severe drinking problem.

The CAGE questionnaire for identifying alcohol problems:

1   Have you ever felt you should Cut down on your drinking?
2   Have people Annoyed you by criticising your drinking?
3   Have you ever felt bad or Guilty about your drinking?
4   Have you ever had a drink first thing in the morning to steady your nerves or to get rid of a hangover (an 'Eye-opener')?

The experiences of people addicted to drugs or alcohol, to the extent of suffering physical withdrawal symptoms when they cannot get the next dose of their chosen substance, may seem very unfamiliar to the majority of people who never develop such a problem with use of either substance. But a degree of dependence on something is almost universal. It may be another person or a pleasant experience rather than a drug, which induces the intense anticipation of repetition and a mixture of psychological and physical symptoms of distress when the object of desire is unavailable.

Epidemiology

We can consider the impact of alcohol and drugs either at the level of the individual or the population to which that individual belongs. The epidemiological approach considers the distribution of patterns of drug use and their consequences in a population. It is useful to distinguish between

- *epidemiology of use*: how patterns of drug or alcohol consumption are distributed through populations;
- *epidemiology of harm*: how problems related to drug or alcohol use are distributed through populations

Alcohol

Epidemiology of use

Information about people's drinking habits comes from a range of sources, including surveys which ask people to remember how much alcohol they have consumed over a defined period of time, such as the preceding week, from official statistics produced by Customs and Excise and based on the amount of alcohol on which duty has been paid. Neither of these can be entirely accurate: people may be reluctant to admit or unable to remember just how much they really drink, while Customs and Excise returns do not include alcohol which is smuggled in, bought duty free or produced at home. It is possible to test for underestimation in self-reported surveys by comparing the consumption of alcohol per head of population predicted from the admitted drinking of the surveyed sample with the per capita consumption derived from Customs and Excise returns for the same period. The Office of Population Censuses and Surveys carried out in 1987 a national survey of drinking habits in England and Wales. This survey estimated that the average annual per capita consumption of pure alcohol was 4.2 litres. Customs and Excise returns for the same year give an average figure of 7.4 litres per year. The reasons for this discrepancy are a tendency for people, especially heavy drinkers, to underestimate their own consumption and for a lower rate of response to surveys from heavy and problem drinkers. Interestingly, people seem to give a more accurate response to questions about drinking habits when a doctor or other health professional is enquiring as part of a clinical consultation. This greater accuracy in the clinical setting may indicate that people perceive it as being more in their own interests to be honest with their doctor than in a population survey where there is no personal gain for the respondent from admitting the true extent of their drinking.

Overall levels of alcohol consumption can be expressed as weight or volume of pure alcohol but are now often described in 'units' of alcohol, since this is a more convenient measure for people to understand and use in everyday life. One 'unit' of alcohol corresponds to 9 grams of pure ethanol. More importantly, a unit matches fairly closely the standard measures in which alcoholic drinks are sold in pubs and bars. So, a unit of alcohol is equivalent to half a pint of normal strength beer, a single measure of spirits or a glass of wine.

A unit of alcohol (equivalent to 10 ml [9 grams] of pure alcohol) is contained in:

- half a pint of beer of normal strength (4 per cent)
- one standard pub measure (one-sixth of a gill) of spirits
- one glass of wine (125 ml)

Data from a number of such population surveys suggest that, in the UK, the vast majority of adults drink alcohol at least occasionally. Only about 5 per cent of men and 8 per cent of women describe themselves as tee-totallers. This average figure conceals a great deal of variation between men and women, different age groups and different communities. Men and women in the 18–24 age group tend to have the highest consumption levels, declining progressively to those in the 65+ age group who have the lowest levels of consumption. Older adults are less likely to be drinkers and there are a number of ethnic and religious groups who maintain a tradition of abstinence. Men usually drink more than women, but there is evidence from recent surveys that the drinking habits of younger women are becoming more similar to those of men.

How much alcohol we drink overall, and in what form, varies between one country and another and across time within each country. Britain has a relatively low level of alcohol consumption compared with most European countries. Historically, average levels of alcohol consumption were high in the eighteenth century and dropped to very low levels during World War II and the 1950s. Since then, consumption has risen quite a bit and we are now drinking about as much as at the beginning of the twentieth century. Economic and social factors play an important part in determining these trends; the high consumption of beer in the eighteenth century was partly enforced by the difficulty of obtaining a supply of clean water fit for drinking. The low levels of consumption in the early part of the twentieth century were largely the result of wartime rationing and a severe economic recession during the 1930s. The Family Spending survey of 1998 showed that spending on alcohol has risen by 40 per cent in real terms over the last 30 years in Britain. This increase has not been evenly distributed across British society: the biggest rise occurred in the poorest fifth of the population, where alcohol spending went up by 80 per cent.

Britain is still predominantly a beer-drinking culture, although the popularity of wine has increased greatly in recent years. This pattern of preference is similar to countries such as Denmark, Germany and Ireland. It contrasts with countries such as France, Italy and Spain, where wine has always been the preferred beverage, and another group of countries such as Poland and Sweden, where spirits are the most popular choice for drinkers.

The pattern of drinking varies between countries as well. Britain shares with Scandinavian countries a pattern of alcohol consumption which has sometimes been described as episodic or binge drinking. People will often go for several days without drinking at all but will then drink a relatively large amount in a single session (e.g. at the weekend). This contrasts with wine-drinking countries, where there is a pattern of steady drinking spread more evenly throughout the week. People may have a small amount to drink with each meal but will not usually consume a large amount in a single session. One very obvious implication of these differences is in the amount of drunkenness. Episodic or binge drinking is more likely to lead to intoxication and drunken behaviour.

How much should we drink?

Current recommendations mostly concentrate on defining what is a sensible upper limit. The UK's medical royal colleges recommend an upper limit of sensible consumption of 21 units per week for men and 14 units per week for women. More recently, the UK Department of Health produced a higher figure of up to 28 units per week for men or up to 21 units per week for women. The fact that recommended levels for women are less reflects both their lower average body weight and their lower proportion of body water compared with men. These upper limits of sensible consumption are somewhat arbitrary since there is no clear threshold level below which drinking is always harm-free and above which it is always harmful. Currently, there is still some controversy over whether there should be a minimum recommended level. In other words, should people be advised to drink some alcohol on health grounds? There is mounting evidence that people who drink no alcohol at all suffer higher rates of heart disease than those who drink a little. This phenomenon has sometimes been referred to as 'the J-shaped curve' or the 'U-shaped curve' because the graph plotting death rate against alcohol consumption shows that both teetotallers and heavy drinkers have higher death rates, particularly from heart disease, than those who drink a little alcohol. We should be cautious about recommending alcohol on health grounds, however, since there is evidence from other countries, which have tried this strategy, that it can lead those who are already drinking too much to drink even more and can cancel out the theoretical benefits from reduced rates of heart disease by increasing the risk of other types of alcohol-related harm, such as accidents.

Epidemiology of harm

Alcohol misuse is associated with an increased risk of a wide range of different problems. It is estimated that alcohol misuse contributes to about 30,000 deaths per year in the UK. Currently in Britain, about 28 per cent of men and about 11 per cent of women are drinking above the recommended levels of sensible consumption. In a recent survey, 54 per cent of men and 28 per cent of women reported experiencing in the previous year the short-term detrimental effects of drinking too much, such as hang-overs.

Some types of alcohol-related harm:

- physical illness (e.g. cirrhosis, high blood pressure, obesity)
- mental illness
- public disorder, violence and crime
- family disputes, relationship break-ups
- child neglect and abuse
- road accidents
- accidents at work and in the home
- fire
- drowning
- absenteeism and unemployment

Some types of alcohol-related harm, such as cirrhosis, are directly related to the pharmacological effects of alcohol on the body. There are other causes of cirrhosis apart from alcohol, but, in Europe and North America, the vast majority of cases of cirrhosis are alcohol related. The frequency of cirrhosis in the population is therefore directly proportional to the average level of alcohol consumption in the population, and is thus often used as a relatively 'pure' marker of alcohol-related harm for the purposes of population studies. In the UK, rates of cirrhosis in young men and women are rising sharply (Donaldson 2001). Other types of harm, such as accidents, violence and crime, are more related to episodes of drunkenness, and hence the pattern of alcohol consumption is important as well as the overall level. The amount and type of alcohol-related harm in a country therefore reflects both the level and pattern of alcohol consumption. Wine-drinking countries such as Italy or France, with a relatively high average level of alcohol consumption but not much binge drinking, often have quite high levels of alcoholic liver disease but relatively low levels of drunkenness and alcohol-related violence. Countries such as the UK, with a lower overall level of alcohol consumption but more of a binge drinking pattern, usually have a lower level of alcoholic liver disease but more of those types of alcohol-related harm which result from drunkenness.

Most of the alcohol-related harm which occurs in the UK population is accounted for, not by alcoholics or heavy drinkers, but by people who have suffered harm as a result of intoxication through drinking too much at one sitting or drinking at the wrong time, such as before driving. Most of these people will have an overall weekly level of alcohol consumption which is within the moderate range. This means that the best way of reducing the total amount of alcohol-related harm in the population would be to get all the moderate drinkers to drink a little less and with more care. Such a change would have a much bigger combined impact on the total amount of alcohol-related harm experienced by the population than if every one of the relatively small numbers of alcoholics became a total abstainer. This apparently curious conclusion is sometimes called 'The Preventive Paradox'. It illustrates the difference between individual risk and population risk, and it also applies to many other types of risk factor for illness, such as high blood pressure. The risk to the individual drinker is greatest for the heaviest drinkers but the risk to the population reflects both the number of people at risk and their individual levels of risk. The much greater number of moderate drinkers more than makes up for their lower levels of individual risk and means that most of the population's risk of alcohol-related harm arises from moderate drinkers.

Not just how much but also when

The overall level of alcohol consumption is not the only factor determining the risk of alcohol-related harm. The pattern and circumstances of drinking also make a difference. Consider someone who drinks 5 pints of beer each week. This amounts to 10 units of alcohol and is therefore well within the recommended weekly levels of sensible drinking. However, if our drinker consumes all 5 pints rapidly in one session on an empty stomach and immediately follows it by driving a motorbike on an icy road, then there is a very high risk of harm from an alcohol-related road accident.

Recommendations on sensible drinking should also include advice on the risk from intoxication under particular circumstances such as driving or operating machinery. Different circumstances demand different advice; many women who enjoy a drink under normal circumstances prefer to avoid alcohol completely during pregnancy rather than run even the smallest risk of alcohol-related harm to the foetus.

What are the factors influencing alcohol consumption?

A wide range of factors influence how much people drink.

Some factors which influence the level of alcohol consumption:

- affordability of drink
- availability (e.g. number of outlets, opening hours)
- cultural standards
- peer pressure
- advertising

The most important of these seems to be how readily available and affordable alcohol is. Putting it very simply, the lower the real price of alcohol and the easier it is to get it, the more people will drink. There is a very convincing relationship between the affordability of alcohol, the average level of consumption in the population and indicators of harm, such as the death rate from cirrhosis (see Figure 1.1).

The notion of availability of alcohol would include such things as how many bars or licensed outlets there are in an area, how much of the day they are open and more intangible things such as how ready people are to offer alcohol at social functions. These can show quite dramatic changes over fairly short periods of time. Over the past 20 years, it has become much less socially acceptable to drink and drive. People are more likely to offer non-alcoholic alternatives alongside alcohol at parties.

Drugs

In general

In terms of the impact at a population level, many of the same considerations apply to drugs as to alcohol. The more easily available and the more

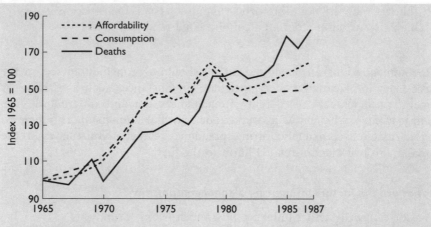

Figure 1.1 Alcohol consumption, affordability of alcohol and deaths from chronic liver disease UK 1965–87.

affordable drugs are, the higher will be the level of consumption and the higher the risk of resulting harm. None of the drugs considered in this chapter are used by anything like the proportion of the population who drink alcohol, but the frequency of use of most drugs is increasing with time and some of the commonest, such as cannabis, are now used by a substantial minority of the population.

There are some effects that are particular to specific drugs; for example, heavy amphetamine use carries a particular risk of paranoid psychosis. Other types of harm relate to the route by which the drug is taken; for example the risk of picking up HIV infection from the use of contaminated injecting equipment. And there are problems for drug misusers which arise from the illegal status of some drugs, particularly those called 'hard drugs' or 'controlled drugs' which are specified in the Misuse of Drugs Act and whose possession or dealing attract heavy penalties (e.g. heroin and cocaine).

Epidemiology of use

We do not have for drug misuse the detailed historical records available for alcohol consumption. This is because some drugs, such as Ecstasy, are relatively new in this country and also because their illegal status means that those who produce, sell or use them may prefer to conceal the fact rather than volunteer the information in response to enquiry. The information that we have is derived mainly from: what people are prepared to admit in confidential, anonymous surveys; what the law enforcement agencies are able to detect; and statistics about those people who come to the attention of health services, either seeking help for their drug habit or with health problems which are obviously related to drug misuse, such as an accidental overdose of heroin. Recent UK surveys suggest that, for the population as a whole, about 6 per cent of people take an illegal drug in any one year (mainly cannabis) and that almost one-third of people aged 16–29 have tried an illegal drug at least once. In 1999, adolescents were especially likely to report some use; for example the European Centre for Monitoring Drugs and Drug Abuse estimates 42 per cent of British 15- to 16-year-olds have tried an illicit drug. So, drug use is much less common than alcohol use but is no longer rare.

The following sections summarize just a few details about some of the main categories of drugs which are misused in the UK. For further details, see the references. Many people who misuse one type of drug also misuse alcohol and other drugs, too.

Controlled drugs (hard drugs)

In the UK, the Misuse of Drugs Act 1971 and its associated regulations control many aspects of the misuse of illicit drugs. These classify what are perceived to be the most dangerous drugs according to their potential for harm. The term 'controlled drugs' is sometimes used to describe collectively this group of drugs, including heroin and cocaine, which are regarded as sufficiently serious threats to require inclusion in legal and official regulations. The popular term 'hard drugs' is often used in a similar way but is less precise because there is no official definition of it.

Controlled drugs specified in the UK Misuse of Drugs Act 1971:

Class A includes cocaine, heroin, morphine, methadone and LSD.
Class B includes oral amphetamines, barbiturates and cannabis.
Class C includes benzodiazepines (minor tranquillizers).

Class A drugs are regarded as being the most dangerous and are subject to the most careful controls and the heaviest penalties for misuse.

Epidemiology of use

The Home Office used to maintain an Addicts Index which recorded some basic details supplied on a confidential basis by doctors who saw patients who were using opiates or cocaine. This system has been superseded by requiring each regional office of the NHS Executive to maintain a drug misuse database. Each regional database now receives anonymized information sent in by all the agencies which see people presenting for treatment of drug misuse. Most people who use drugs will not be in contact with a doctor or any other service, and so these figures very much represent the tip of the iceberg; it has been estimated that the drug users known to treatment agencies are only about 10 per cent of all the drug users in the population at any one time. In summary, these two sources show that the number of known drug misusers has increased greatly over the past 30 years and is still increasing. The Home Office Addicts Index recorded just over 2,000 notifications in 1981 but over 20,000 in 1991. Between 1990 and 1993, there was a 30 per cent increase in the number of people notified as using opiates and a 130 per cent increase in the number of people notified as using cocaine. The predominant drugs of misuse among people notified to the Addicts Index and the regional drug databases are the opiates, particularly heroin. There has been a definite increase in the misuse of cocaine in recent years but it has not as yet reached anything like the epidemic proportions that were predicted from the experience in some US cities in the mid-1980s. Newer drugs such as Ecstasy have emerged, amphetamines remain popular and there has been some crossover from drugs which were originally prescribed by doctors for medical use but which have found their way onto the streets where they are misused, such as temazepam, a minor tranquillizer.

There is great interest in the proportion of drug users who take their drugs by injecting them. This is because injection using contaminated equipment carries major additional risks over and above the risks of the drug itself, principally from infection with viruses such as hepatitis and HIV. The proportion of notified addicts who report injecting declined from about 60 per cent in 1990 to just under 50 per cent in 1995. The proportion of drug addicts who are known to be infected with HIV has been estimated at between 1 per cent and 5 per cent across the UK as a whole, but with much higher levels in some localized areas at certain times, such as Edinburgh in the early 1980s when more than half of local heroin users were thought to be HIV positive.

Epidemiology of harm

There are ranges of population markers of drug-related harm.

Examples of drug-related harm:

- accidental overdose
- infections: hepatitis B and C, HIV, abscesses at injection sites
- mental illness: psychosis, depression
- suicide

- involvement in crime to finance a drug habit
- injury from accidents or violence
- legal, social and financial problems

Indicators or markers of drug-related harm include the number of deaths from causes known to be related to drug misuse, the number of injecting drug misusers who become infected with HIV, the numbers of arrests for drug-related offences and the financial cost to the country of drug-related crime. There are problems of classification and interpretation with all of them: it can be quite difficult to decide whether the death of a known drug user is actually due to their use of drugs, and, conversely, many deaths which are actually drug related may not be noticed as such if the link is indirect (e.g. a road accident). How are we to know if a sexually active injecting drug user, who is discovered to be HIV positive, acquired the infection through use of contaminated injecting equipment or by unprotected sex? The number of arrests which the police make for drug-related offences depends not only on how many people are committing such offences but also on how much resources the police decide to devote to this area of law enforcement when there are many other calls on their time.

In England and Wales for the period 1992–93, these markers of drug-related harm showed that there were about 1,200 drug-related deaths, mostly from overdose but about fifty from AIDS. About 68,000 people were convicted of a drug-related offence and the cost of thefts related to heroin use was estimated at £864 million. Acquisitive crime involving drug users has continued to rise through the 1990s, and Business in the Community (1999) now estimate that 66 per cent of retail crime is drug related.

Cannabis

Cannabis is by far the most widely used illicit drug in the UK. Survey figures suggest that about 21 per cent of the adult population have tried it at least once and that perhaps 10 per cent of teenagers report frequent use. It accounts for more than 75 per cent of arrests for possession of drugs. The drug is usually smoked as a mixture of tobacco with either the dried leaves or the resin of the plant. The extent to which it causes harm is a matter of both scientific and public controversy. Some people regard it as a harmless natural herb less dangerous than alcohol; others see it as a dangerous 'gateway drug' which leads users on to experiment with hard drugs. Most moderate users of cannabis experience no serious problems. Some users have short-lived attacks of anxiety or even psychotic reactions, but long-standing mental illness seems to be rare and it is not certain that it can be attributed to the drug. Like alcohol, cannabis does impair performance on skilled tasks such as driving and so may increase the risk of road accidents in those who use it. It persists in the body for much longer than alcohol, several weeks after use, and so may exert effects for quite a long time.

Volatile substances

Volatile substance abuse seems to be the most correct term for what is sometimes called 'glue-sniffing' or 'solvent abuse', because all the substances employed in this way are volatile compounds which vaporize readily but not all of them are either glue or solvents. The range of substances includes butane gas, petrol, thinners, dry-cleaning liquid, adhesives and a range of other products as diverse as fire extinguishers, aerosols, marker pens and typewriter correction fluid. Most of the products are cheap and in common household use, so control is difficult. Legislation attempts to restrict their sale to children but is difficult to enforce. All these volatile substances are inhaled as vapours to produce a state of intoxication. Depending on the nature of the substance used this may involve breathing it in from a plastic bag, squirting it directly into the mouth or soaking a rag which is then sniffed.

Epidemiology of use

Users of volatile substances are distinctly younger on average than most users of other drugs. Surveys of secondary schools have found that perhaps 20 per cent of children report having tried inhaling volatile substances while less than 1 per cent go on to more regular use. It is rare to find anyone still doing it in adult life. There are sometimes reports of 'epidemics' of use in a particular school or youth organization where, for a period of time, a particular type of volatile substance abuse will be popular and then will die out.

Epidemiology of harm

The most serious risks are of sudden death during inhalation either from the direct effects of the volatile substance itself or from accidental injury while intoxicated. There were about 150 deaths of this sort in the UK in 1990. Users have died or been seriously injured as a result of falls, drowning, burns and asphyxiation. Although as many girls as boys admit to sniffing, most of those who die are boys. The use of butane gas in cigarette lighter refills is particularly associated with deaths among young people aged 15–19 (Department of Health 1999). Worryingly, many of the deaths appear to involve novice users in their first attempts. Regular users may experience skin rashes from contact with the chemicals involved, and there have been reports of long-term damage to the liver, kidneys and nervous system.

Prescribed drugs

Many drugs that can legitimately be prescribed by doctors for medical reasons also have a potential for abuse or dependence. This may involve deliberate diversion of legally prescribed drugs for illicit sale on the black market, intentional abuse by taking them in a way other than that prescribed or unintentional dependence through long, continued use. The drugs most commonly involved in this way are minor tranquillizers such as the benzodiazepine drug temazepam, appetite suppressants, sleeping tablets and opiate-like drugs prescribed as painkillers.

Epidemiology of use

Some of those who misuse prescribed drugs are the same people who are also misusing illicitly produced drugs such as heroin or cocaine. It was popular for a time to extract the liquid contents of temazepam capsules and inject them. Appetite suppressants have a mild stimulant effect and may be used as a substitute for other more powerful stimulants such as amphetamines or cocaine. There is another quite different population of people who become unintentionally dependent on a prescribed drug as a result of prolonged prescription. Women and older people are more likely to be on long-term prescription of minor tranquillizers. This group may include a number of older people who began taking the medication years ago before the dangers of causing dependence were fully realized. Many of these people may be unaware that they are physically dependent until they try to stop the tablets and experience withdrawal effects. In Britain, the peak of benzodiazepine prescribing was in the 1970s and it had declined by about a third by 1990.

Epidemiology of harm

Deliberate misuse of prescribed drugs is often part of a pattern, which involves misuse of other illicit drugs, and it can be difficult to decide which of the misused drugs is responsible for any resulting harm. Serious problems can result from injecting drugs which were never intended to be taken in this way and which may, for example, solidify inside the veins, blocking the circulation of blood. The diversion of prescribed drugs to the black market increases the availability of drugs for misuse and so increases the risk of encouraging drug problems.

Patients who become unknowingly dependent on prescribed minor tranquillizers may experience problems as a result of being under long-term sedation. They may be more likely to have an accident at work or while driving and more inclined to tolerate life problems such as an unsatisfactory relationship which they might otherwise get round to resolving. Elderly people on long-term prescriptions of minor tranquillizers may be more likely to become confused or suffer a fall. Concerns about accidents such as falls are growing more widespread as older people increasingly combine the use of alcohol with their prescribed minor tranquillizers (Leach 1999).

Bibliography

Business in the Community (1999) *Drugs – the Business Agenda*. London: UK Anti-Drugs Co-ordinating Unit.

Butler, D. (1998) 'Dirty money', *Accountancy*, **122**, November, 34–6.

Chick, J. and Cantwell, R. (eds) (1994) *Seminars in Alcohol and Drug Misuse* (Royal College of Psychiatrists). London: Gaskell.

Day, N. E. (1993) *The incidence and prevalence of AIDS and other severe HIV disease in England and Wales for 1992–1997: projections using data to the end of 1992.* (Report of a Working Group chaired by Professor NE Day), *Communicable Disease Report*, **3** (suppl. 1), S10.

Department of Health (1989) *AIDS and drug misuse* (Report by the Advisory Council on the Misuse of Drugs, Parts 1 and 2). London: Her Majesty's Stationery Office.

Department of Health (1992) Drug and Solvent Misuse: A Basic Briefing. London: Department of Health.

Department of Health (1999) 'Curbing use of lighter refills', *Target*, **35**, 22.

Donaldson, L. (2001) *Liver Cirrhosis – Starting to Strike at Younger Ages* (Annual Report of the Chief Medical Officer 2001). London: Department of Health.

Edwards, G. (1996) 'Sensible drinking', *BMJ*, **312**, 1.

Faculty of Public Health Medicine (Royal Colleges of Physicians) (1991) *Alcohol and the Public Health. The Prevention of Harm Related to the Use of Alcohol*. London: Macmillan.

HM Government (1995) *Tackling Drugs Together. A Strategy for England 1995–1998*. London: Her Majesty's Stationery Office.

Interdepartmental Working Group (1995) *Sensible Drinking*. London: Department of Health.

Leach, E. (1999) Bottling up problems. *Nursing Times*, **95**(8), 18.

McElduff, P. and Dobson, A. J. (1997) 'How much alcohol and how often?' (Population based case-control study of alcohol consumption and risk of a major coronary event), *BMJ*, **314**, 1159–64.

Webster, P. (1999) 'France counts cost of alcohol and tobacco culture', *Guardian*, 26 October, p. 15.

Chapter 2

Systemic effects of excess alcohol consumption

Alexander Baldacchino

Key points

Detection of a previously unrecognized drink problem can provide an opportunity to influence what otherwise is a major cause for considerable suffering and the tragedy of many premature deaths. The figures below give us an introductory impression of this extent:

- Approximately one-fifth of men, admitted to an acute general medical ward, have alcohol-related problems.
- It will cost the health service £150 million a year to treat.
- About one-half of head injuries and one-third of all patients seen in casualty have a high blood alcohol level.
- There are at least 8 million heavy drinkers in the general population with 10 per cent of these people exhibiting alcohol dependency.
- The overall mortality amongst a group of alcoholics is twice that found in the general population. The principal causes are accidents, suicide, cirrhosis, heart disease and lung cancer.
- Surveys suggest that only 10 per cent of all heavy drinkers registered with a general practitioner are diagnosed as such.

Alcohol intoxication

There are over fifty words in the English language for being drunk. Drunkenness is one of the richest areas of common experience, pleasures and problems. Authorities would agree that physiological dysfunction occurs for most people at blood alcohol concentration in excess of 30 mg per 100 ml (30 mg %), the equivalent of drinking a pint of beer or two glasses of wine or spirits. Drinking each unit of ethanol per hour increases the blood alcohol concentration (BAC) by 15 mg % (Table 2.1).

Table 2.1 Acute intoxication

BAC mg/100 ml	Effects for drinkers with average tolerance
20	Increased sense of well-being and reduced reaction time
40	Mild disinhibition
60	Mild impairment of judgement and decision making
80	Reduced physical co-ordination
100	Reduced social and physical control
150	Observable intoxication with amnesic episodes
350	Incontinence and sleepiness
500+	Coma, breathing difficulties and death due to acute alcohol poisoning

BAC = blood alcohol concentration.

Physical harm due to alcohol consumption

There are two aspects of alcohol-related physical disabilities. These are harmful effects directly caused by acute alcohol intoxication and problems due to regular heavy alcohol consumption. There is considerable overlap between the two.

Harmful effects secondary to acute alcohol intoxication

Most of the medical problems (Table 2.2) arising from intoxication come about either due to unco-ordination or due to the acute toxic effects of alcohol on various organs. Over 35 per cent of drivers in Britain killed in road traffic accidents have a blood alcohol concentration of over 80 mg %. Young children, who may ingest alcohol by choice, by accident or under duress from adults, are especially vulnerable both to fatal alcohol poisoning and to long-term neurological damage if they survive (Vale 1999).

Table 2.2 Harmful effects secondary to acute alcohol intoxication

Acute alcohol poisoning and overdosage
Amnesic episodes
Drug overdoses and failure to take prescribed medicine
Suicidal behaviour
Acute inflammation of gut lining (acute gastritis)
Acute pancreatitis
Low blood glucose
Hypothermia
Trauma (e.g. head injury and shoulder dislocation)
Epilepsy
Damage to unborn baby
Gout
Disturbances in cardiac rhythms (arrhythmias)

The overall mortality and morbidity rates directly related to acute intoxication are unmeasurable due to unrecorded accidents that occur in the workplace (e.g. the use of heavy machinery and a disregard for safety procedures in building sites when drunk) and at home (e.g. domestic violence and abuse, improper child care). However, one measurable pattern is now emerging for a grave consequence of acute intoxication – ischaemic strokes in adults aged less than 40 years within 24 hours of their starting an alcohol 'binge' (Puddey et al. 1999). A possible association has also been reported between maternal drinking after a birth and increased rates of Sudden Infant Death Syndrome (Legge, 1999).

Harmful effects due to regular excessive consumption of alcohol

Alcohol can damage nearly every organ and system of the body and leads to premature death (Table 2.3). Alcohol interacts with a wide variety of prescribed medication. Acute alcohol intake prolongs the effects of prescribed benzodiazepines, antipsychotic and antidepressant drugs (Table 2.4). In regular heavy drinkers, the metabolism of anticonvulsants, benzodiazepines and warfarin is increased due to the induction of liver enzymes. If liver disease develops then the metabolism of many drugs will be reduced.

Risk and level of consumption

Heavy drinkers have higher rates of morbidity and mortality than light drinkers. The exact nature of this association for each harmful consequence is less clear-cut.

The WHO International Agency for Research on Cancer concluded in 1987 that alcohol is carcinogenic. Coates et al. (1984) described a linear relationship between excessive alcohol consumption and increased risk of developing cancer. Subsequent studies concluded that the increased risk is not applicable to all types and location of cancer formation. Excessive alcohol consumption increases the risk of malignancy of the tongue, mouth, tonsils, larynx, liver, pancreas, lung, breast and cervix, but is not linked to malignant melanoma of the skin and malignancies of the bladder, prostate, stomach and ovary (Sigvardsson et al. 1996; Longnecker 1995). This multiplicity of site-specific malignancies is a result of ethanol potentiating the carcinogenic risk associated with certain agents present in the environment. Ethanol itself is not a tumorigenic agent. Possible pathways are that ethanol induces enzymes like P450 2EI present in the liver or enhances exposure of nitrosamines in other tissues (Anderson et al. 1995). Free radicals (e.g. lipid peroxide products released during the oxidation process of ethanol to acetaldehyde) have been directly related to oesophageal malignancies (Eskelson et al. 1993).

Table 2.3 Harmful effects due to regular excessive consumption of alcohol

Organs/Systems	Medical conditions with alcohol as an etiologic component
Major hazards	
Respiratory	Tuberculosis; pneumonia; lung abscess; bronchiectasis; cancer of the pharynx and lungs
Gastrointestinal	Oesophagitis and gastritis; peptic ulceration; tearing of lower oesophagus; cancer of mouth, tongue, pharynx, oesophagus, colon
Liver and pancreas	Hepatitis; fatty liver; cirrhosis; cancer of the liver; acute and chronic pancreatitis
Cardiovascular	Arrhythmias; coronary heart disease; cardiomyopathy; hypertension
Central nervous system	Wernike–Korsakoff syndrome; dementia; cerebellar, pons and brain stem degeneration
Minor hazards	
Bone and muscles	Gout; degeneration of bones; myopathies
Blood	Anaemia secondary to bleeding; liver disease; malnutrition; chronic infections and direct toxic effect of alcohol on bone marrow
Peripheral nervous system	Sensory and motor neuropathies
Endocrine	Cushingoid syndrome with high cortisol levels; feminization; diabetic symptoms due to low and high blood glucose levels
Sexual functions	Impotence; testicular atrophy with low sperm count; loss of sexual drive; amenorrhoea and anovulation
Nutritional	Vitamin and mineral deficiency (e.g. thiamine deficiency causing neuropathies and CNS effects)
Psychiatric	Depression; anxiety; panic attacks and sleep disorders
Pregnancy	Foetal alcohol syndrome*; increased risk of spontaneous abortion, still births and congenital malformations; intra-uterine growth retardation
Skin	Rhinophyma ('red nose'); psoriasis, acne, rosacea

* *Foetal Alcohol Syndrome (FAS)* is a constellation of features that include central nervous dysfunction, abnormal facial features, behavioural problems and growth deficiency. This can lead to learning disabilities and hyperactivity in later years. This syndrome is significantly more likely to happen in pregnant mothers with a daily average consumption of 10 units during the first trimester. The risks are higher if the mother also smokes and uses illicit substances. The incidence is very low and there is evidence that lower levels of alcohol intake harm the foetus giving a spectrum of Foetal Alcohol Effects (FAE) that are dose related. Children born to mothers drinking more than 2 units daily have lower birth weight and facial dysmorphic features.

Cryer *et al.* (1999) also found reduced use of 'preventive' services by heavy drinkers; for example, women who drink excessively participate less than other women in the breast and cervical screening programmes, and men with heavy drinking are less likely to visit a dentist (who might detect oral cancers at an early stage).

Anderson *et al.* (1993) established a threshold level between drinking and cirrhosis. Chronic alcohol consumption is the commonest cause of liver

Table 2.4 Interaction of alcohol with prescribed drugs

Medicine commonly used	Interaction with alcohol
Analgesia (e.g. opiates [morphine])	Increased sedative and hypotensive effects of opiates
Antidepressants	Increased sedative effects of tricyclic antidepressants; hypertensive crisis with monoamine oxidase inhibitors (MAOI) antidepressants especially with Chianti wine
Anxiolytics	Increased sedative effect of anxiolytics (e.g. benzodiazepines)
Antidiabetic agents	Increased hypoglycaemic (low glucose) effect
Antiepileptic agents	Increased, in central nervous system, side effects of carbamazepine
Antihypertensive agents	Increased hypotensive (low blood pressure) effect
Antibiotics	Disulfiram-like reaction with metronidazole, tinidazole and cephamandole

damage in the UK, associated with 75 per cent of all cases of cirrhosis. Fatty liver is present in 90 per cent of chronic alcohol drinkers but only 8–30 per cent of these drinkers actually develop cirrhosis. Fillmore and colleagues (1998) have developed a mathematical model to explain how a small increase in drinking across this chronically 'at risk' population can rapidly push up the percentage developing cirrhosis.

Factors that increase ethanol oxidation to acetaldehyde or reduce acetaldehyde clearance will result in increased acetaldehyde levels in the liver. This will result in a toxic derangement of vital hepatic cell function and subsequent cell swelling. The ALDH2 (a mitochondrial isoenzyme of aldehyde dehydrogenase) is the most important liver enzyme that is responsible for the proper breakdown of acetaldehyde. Oriental populations have a genetic predisposition to possess an active form of this isoenzyme, hence protecting them from developing liver damage.

Marmot and Brunner (1991) and Doll et al. (1994) determined a U-shaped curved relationship between coronary heart disease amongst moderate alcohol drinkers. Chronic, moderate to high alcohol consumption increases the plasma high density lipoprotein type 2 (HDL2) fraction and thus may be protective against coronary heart disease. However, these beneficial effects of alcohol are weakened by the increasing tendency for drinkers to be obese, to be hypertensive and to develop other heart problems notably alcoholic heart muscle disease (AHMD). The latter is characterized by an enlarged heart, dilated left ventricle and irregularities of cardiac conduction, with functional and biochemical abnormalities in heart muscle. Yet again, acetaldehyde is implicated in the toxic process of heart muscle damage with the help of superoxide radicals (Hess et al. 1981).

We must take into account other variables that might enhance the risk of a drinker developing medical problems that are consequential to excessive alcohol consumption. These include:

1   *Pattern of drinking.* Is it binge drinking, regular or weekend consumption?
2   *Individual susceptibility to alcohol.* This includes:
    •   Gender differences: Females tend to have a smaller volume of distribution of ethanol because their bodies have a lower water content. Frezza *et al.* (1990) found that the bioavailability of ethanol is much greater in women than in men because women have less gastric first-pass metabolism of ethanol making them more vulnerable to both acute and chronic complications of excessive alcohol consumption.
    •   Ethnicity: The availability of an active form of ALDH2 in oriental populations gives them a protective mechanism not to develop alcohol liver disease; the same protection is not available in Caucasian populations.
    •   Other individual variables such as age, personality profile, genetic susceptibilities and pregnancy.
3   *Unrelated medical conditions.* Examples include a raised cholesterol level, hypertension, obesity, chronic obstructive airway disease.
4   *Use of other intoxicants.* Smoking increases the risk of lung cancer. The risk of overdosage in intravenous heroin users is increased.
5   *Environmental issues.* The social class, isolation and disruption in the drinker's immediate environment. Unemployment and other occupational considerations (e.g. being a bartender or oil rig worker).
6   *The type of beverage consumed.* Spirits are directly associated with increased rates of oesophagus and oral cavity malignancies. Consumption of large amounts of cheap wine is more likely to produce cirrhosis.

The recommended guidelines as advised by the royal medical colleges in the UK (Royal College of Psychiatrists 1987) are found in Table 2.5.

Overall, across a wide spectrum of social environments, a history of alcohol abuse before the age of 50 years doubles the likelihood of men having a physical disability at the age of 75. About 70 per cent of all these problem drinkers are disabled when 75 years old (Vaillant 1999).

Case study

A 45-year-old man was brought to the casualty department by the police. He was described as behaving in a confused and agitated manner. He ordered strangers walking in a nearby park not to approach him and

Table 2.5 Recommended guidelines

	Men (weekly consumption, units)	Women (weekly consumption, units)
Sensible drinking	Up to 21	Up to 14
Risky drinking with increased hazards	22–50	15–35
Harmful drinking with definite health problems	Over 50	Over 35

accused them of plotting against him. He told a nearby taxi driver that his taxi was crawling with ants.

This man used to work as a pub manager but was sacked a few months before this incident because he was constantly drunk at work. He subsequently became homeless and disappeared from the local scene. But a week ago, he was picked up by the police and examined by the police surgeon who thought the man looked slightly jaundiced (yellowish complexion) and tired but that his behaviour did not justify further investigation.

On examination in the casualty department, he was dishevelled, fidgety, restless and had a smell of vomitus. He appeared to be visually hallucinating and occasionally scratching various parts of his body. The man complained of 'insects crawling over my entire body and biting me'. He denied feeling unwell but, on further examination, he was disorientated in time, place and person. His concentration and attention span were also limited.

On physical examination, he had mild to moderate jaundice with an enlarged and tender liver. There was also a peripheral tremor and motor unco-ordination. Thirty minutes after coming into the casualty department, he suffered a grand-mal epileptic fit that needed urgent treatment including medication.

The only information available that might be relevant included a recent prescription of analgesic drugs for stomach aches due to 'liver inflammation'. He denied drinking alcohol to excess.

1 What questions would you ask the man to get a proper drinking history?
2 What do you make of his references to ants and biting insects?
3 What are the possible medical conditions that might explain his presentation at the casualty department?
4 What are the physical consequences of chronic alcohol abuse?
5 What age group most commonly exhibits dependence on alcohol?

Hints: Table 3.10 lists a number of possible screening questionnaires and physical investigations (e.g. liver function tests) which might help to assess the role of alcohol in this person's admission. A number of this patient's problems (e.g. agitation, tremor, vomiting and fitting) *could* result from alcohol withdrawal in someone dependent on this drug. However, in one respect, this case is not at all typical of British alcoholics seen in a hospital casualty department: alcohol dependence is most common between the ages of 20–29 years.

References

Anderson, L. M., Chabra, S. K., Nerukar, P. V., Souliotis, V. L. and Kyrtoppoulos, S. A. (1995) 'Alcohol-related cancer risk: A toxicokinetic hypothesis', *Alcohol*, **12**, 97–104.

Anderson, P., Cremona, A., Paton, A., Turner, C., Wallace, P. (1993) 'The risk of alcohol', *Addiction*, **88**, 1493–1508.

Coates, R. A., Halliday, M. C., Rankin, J. C., Feinmann, S. V. and Fisher, M. M. (1984) 'A case control study of the risk of cirrhosis of the liver in relation to alcohol consumption' (Abstract 34 for the 35th Annual Meeting for the American Association for the Study of Liver Disease, Chicago, 10 November 1984), *Hepatology*, **4**, 1015.

Cryer, P. C., Jenkins, L. M., Cook, A. C., Ditchburn, J. S., Harris, C. K., Davis, A. R. and Peters, T. J. (1999) 'The use of acute and preventative medical services by a general population: Relationship to alcohol consumption', *Addiction*, **94**, 1523–32.

Doll, R., Petro, R., Hall, E., Wheatley, K. and Gray, R. (1994) 'Mortality in relation to consumption of alcohol: 13 years' observation on male British doctors', *Brit. Med. J.*, **309**, 911–18.

Eskelson, C. D., Odeleye, O. E., Watson, R. R., Earnest, D. L. and Mufti, S. I. (1993) 'Modulation of cancer growth by Vitamin E and alcohol', *Alcohol and Alcoholism*, **28**(1), 117–25.

Fillmore, K. M., Roizen, R., Bostrom, A. and Kerr, W. (1998) 'Musing cirrhosis', paper given at the Society for the Study of Addictions, 5 November, York, UK.

Frezza, M., di Padova, C., Pozzato, G., Terpin, M., Baraona, E. and Lieber, C. S. (1990) High blood alcohol levels in women: The role of decreased gastric dehydrogenase activity and first-pass metabolism. *N. Engl. J. Med.*, **322**, 95–9.

Hess, M. L., Okabe, E. and Kontos, H. A. (1981) 'Proton and free oxygen radical interaction with the calcium transport system of cardiac sarcoplasmic reticulum', *J. Mol. Cell Cardiol.*, **13**, 767–72.

Legge, A. (1999) 'Boozing after birth is a risk factor for SIDS', *Nursing Times*, **95** (30), 39.

Longnecker, M. P. (1995) Alcohol consumption and risk of cancer in humans: An overview. *Alcohol*, **12**(2), 87–96.

Marmot, M and Brunner, E. (1991) 'Alcohol and cardio-vascular disease: The status of the U-shaped curve', *Brit. Med. J.*, **303**, 565–8.

Puddey, I. B., Rakic, V., Dimmitt, S. B. and Beilin, L. J. (1999) 'Influence of pattern of drinking on cardiovascular disease and cardiovascular risk factors – a review', *Addiction*, **94**, 649–63.

Royal College of Psychiatrists (1987) *New report of a special committee of the Royal College of Psychiatrists* (p. 178). London: Tavistock Publications.

Sigvardsson, S., Hardell, L., Przybeck, T. R. and Cloninger, R. (1996) Increased cancer risk among Swedish female alcoholics. *Epidemiology*, **7**(2), 140–3.

Vale, A. (1999) 'Alcohol intoxication and alcohol-drug interactions', *Medicine*, **27** (2), 5–7.

Vaillant, G. E. (1999) 'Lessons learned from living', *Scientific American Presents*, **10** (2), 32–7.

Chapter 3

Alcohol and genetic predisposition

Timothy J. Peters and Victor R. Preedy

Key points

- *Alcohol misuse is familial with up to 50 per cent of patients having an affected first-degree relative.*
- *Clinical studies indicate a genetic component (approx. 25 per cent) in the aetiology of alcohol misuse, but there is wide individual variation.*
- *Attempts to identify a gene defect associated with alcohol misuse have, with the notable exception of the highly protective effect of inactive mutant acetaldehyde dehydrogenase, been controversial and unconfirmed.*
- *There is some evidence for a genetic component to the specific end organ susceptibility of individual alcoholics.*
- *Laboratory investigations are useful confirmatory indicators and monitors of increased alcohol intake and organ damage.*

Alcohol misuse has major medical, psychosocial and economic costs for our society. Every aspect of our behaviour is adversely affected by excess consumption and thus knowledge as to its cause is of major importance. The relative contribution of genes and environment in the predisposition to alcohol abuse, the nature vs nurture debate, has been argued for several decades. Plutarch stated that 'drunkards beget drunkards' and, during the nineteenth and early parts of the twentieth centuries, it was assumed that heredity played the major role in the predisposition to alcoholism. Hence, the attempts at prohibition, absolute temperance and the underlying tenets of Alcoholics Anonymous. In Nazi Germany, in the belief that alcohol abuse was genetically determined, alcoholics were sterilized or worse, in an attempt to eradicate alcoholism. However, during the 1950s and later, scientific methods have been applied to the nature–nurture question.

Clinical genetics (Table 3.1)

Early studies documented the high prevalence of alcoholism in first- and second-degree relatives of alcoholics, comparing the rates with those in

Table 3.1 Summary of evidence for genetic basis of alcoholism

Family studies	Alcoholics have a one in three prevalence of having an alcoholic parent. Prevalence much lower in relatives of non-alcoholics
Twin studies	Higher concordance for alcoholism in monozygotic (identical) compared with dizygotic (non-identical) twins
Half-sibling	Two out of three alcoholic half-siblings have an alcoholic biological parent compared with one out of five non-alcoholics
Adoption studies	18 per cent incidence in adoptees with a paternal history of alcoholism compared with 5 per cent of those without. Adopted-out sons of alcoholics have the same increased risk (three- to fourfold) as brothers raised by alcoholic biological parents

Table 3.2 Frequency of paternal and maternal alcoholism in alcoholics and psychiatric and non-psychiatric patients (from Cotton 1979)

Patient group	Paternal alcoholism	Maternal alcoholism
Alcoholics		
Per cent	27.0	4.9
Number of subjects	4,329	3,500
Studies	32	23
Schizophrenics		
Per cent	9.2	0.4
Number of subjects	654	553
Studies	3	2
Psychiatric patients		
Per cent	5.2	1.2
Number of subjects	788	692
Studies	5	3

normal subjects and in patients with other psychiatric disorders. These findings were well reviewed by Cotton (1979) and are summarized in Table 3.2. It is clear that between 20 per cent and 50 per cent of alcoholics have an affected father in contrast to control patients where the figure is less than 10 per cent. Similar conclusions relate to maternal alcoholism, although the percentages are consistently lower. These findings accord with contemporary clinical practice that up to a half of all patient referrals having an affected first-degree relative. There is also an interesting and, as yet unexplained, observation that female patients tend to have an affected maternal relative whereas males tend to have an affected paternal relative. These observations, however, do not distinguish between the relative contributions of genetic and environmental influences and these issues have been addressed by twin, adoption and half sibling studies (Goodwin 1981).

Table 3.3 Genetic influences in behavioural disorders
        (from Plomin et al. 1994)

Disorder	Concordance		Ratio
	MZ	DZ	
Alzheimer's disease	0.60	0.38	1.6
Alcoholism (males)	0.40	0.20	2.0
Ischaemic heart disease	0.29	0.18	1.6
Schizophrenia	0.48	0.16	3.0
Affective disorder	0.67	0.22	3.0
Autism	0.64	0.08	8.0

MZ: monozygotic (identical) twins.
DZ: dizygotic (non-identical) twins.

Table 3.4 Characteristics of two types of alcoholic

	Type 1	Type 2
Usual age of onset	>25 years	< 25 years
Inability to abstain	+	++
Psychological dependence	++	+
Harm avoidance	++	+
Environmental component	++	+
Genetic component	+	++
Gender	Male and female	Males predominate

Comparison of concordance rates for a particular disease in monozygotic (MZ, identical) compared with dizygotic (DZ, non-identical) twins can indicate the genetic contributions to a particular disease. This approach has been applied to a variety of diseases (Plomin et al. 1994) and some of the results are summarized in Table 3.3. MZ/DZ ratios greater than 1 are indicative of a significant genetic component, and it is clear that a positive result is obtained for alcoholism, but the ratio is modest compared with that of several other disorders. The genetic component appears to be similar to that for ischaemic heart disease (1.6), but much less than that for autism (8.0). There is, nevertheless, a significant genetic component for alcohol misuse, although the level of genetic influence varies considerably from patient to patient. On the basis of twin studies, Cloninger (1987) has postulated two types of alcoholics (Table 3.4). Type 2 alcoholics have a major genetic component in their aetiology, and this group is more difficult to treat and is associated with a poorer clinical outcome. It is these patients who should receive the specialist psychiatric care. Patients with Type 1 alcoholism may respond to straightforward medical advice and counselling. This classification thus has direct value in the clinical care of alcoholics.

Table 3.5 Genetic defect in Oriental acetaldehyde dehydrogenase

	Active	Inactive
DNA (Exon 12)	AAA	GAA
Protein (residue 487)	Glu$^-$	Lys$^\oplus$

Interesting as twin studies are, they can be criticized on such features as birth trauma, special relationships and parental influences. Cross-adoption studies offer an alternative approach. These studies show that, at least for males, the genetic make up of an individual is an important determinant of whether they develop alcoholism (Goodwin 1981). Similarly, studies of half-siblings have provided further evidence in favour of a genetic component in alcoholism.

Animal studies have emphasized that genetic elements are important in both the preference for alcohol and its toxic consequences (Li and McBride 1995). Several independent strains of alcohol-preferring rats have been bred. This has allowed the identification of various neurotransmitters and neuronal pathways implicated in the development and maintenance of alcoholism; for example, in one strain of alcohol-preferring rats (Badia-Elder et al. 1999), their characteristically high intake of alcohol is prevented by drugs which block, selectively, the μ-opioid receptor in the brain. These findings have led to the increasing use of chemotherapeutic approaches to the treatment of alcoholism (e.g. opioid antagonists such as naltrexone or serotonin-uptake inhibitors such as fluoxetine). Drugs interfering with dopamine and GABA pathways may also have a role in the treatment of alcoholism (Chick 1996); for example, inbred colonies of vervet monkeys can produce some strains that have an abnormality in dopamine synthesis, which show steady heavy drinking behaviour when alcohol is available (Wilson 1999).

A clear example of the genetic influence on the development of alcoholism is found in the Oriental flush reaction (Shibuya and Yoshida 1988). These individuals have a base change in the gene for acetaldehyde dehydrogenase 2, the mitochondrial enzyme responsible for most of the metabolism of acetaldehyde to acetate (Table 3.5). This base change affects the charge on the individual polypeptides of the enzyme tetramer. The tetramer can no longer be assembled and the enzyme is largely inactive, although enzyme protein can be readily identified immunologically in these subjects. These individuals, like subjects taking the drug disulfiram, a potent inhibitor of mitochondrial acetaldehyde dehydrogenase, can readily metabolize ethanol to acetaldehyde but are unable to further oxidize the acetaldehyde to acetate. High levels of acetaldehyde are clearly aversive to the further consumption of ethanol and, hence, the use of disulfiram in the prevention of relapse in alcoholics. High levels of acetaldehyde cause nausea, abdominal

Table 3.6 ALDH polymorphism in Japanese (percentage distribution of ALDH variants; Higuchi et al. 1996)

Genotype	Alcoholics (%)			General Population (%)
	1979	1986	1992	
Asian variant				
Homozygous	0	0	0	10
Heterozygous	3	8	13	42
Western variant				
Homozygous	97	92	87	48

discomfort, hypotension and intense unpleasant flushing reactions. Approximately 50 per cent of Orientals have this defective enzyme and these individuals are relatively (approx. fivefold) protected against developing alcohol abuse. This is particularly evident for individuals who are homozygous for the abnormal variant. Thus, most Japanese and Chinese alcoholics will have the active 'Western form' of acetaldehyde dehydrogenase (Table 3.6).

Certain European individuals, particularly females, exhibit a mild degree of flushing after consuming modest amounts of alcohol. Some of these individuals have an abnormal cytosolic acetaldehyde dehydrogenase (Yoshida et al. 1989). This enzyme is present in a variety of tissues including erythrocytes, but plays a relatively minor role in systemic acetaldehyde metabolism. These individuals do not have raised plasma acetaldehyde levels after drinking alcohol and do not show an aversion to continued drinking (Whitfield 1997).

Hepatic alcohol dehydrogenase genes show a marked degree of polymorphism with a wide variation between different racial groups. In addition, these isoenzymes show markedly different kinetic properties. However, data on the associations between enzymic polymorphism and the development of alcoholism remain controversial (Day and Bassendine 1992; Ferguson and Goldberg 1997). Within the Collaborative Study on the Genetics of Alcoholism, one attempt to improve the investigation of enzymes like alcohol dehydrogenase (Williams et al. 1999) combines clinical measures using multiple diagnostic systems and neurophysiological measures in the same subjects.

A number of other candidate gene polymorphisms have been implicated in the susceptibility to alcohol dependence. Particular interest has been paid to a polymorphism of the dopamine D2 receptor. An early report claimed an association between the A1 allele and alcoholism (Blum et al. 1990). A successive flurry of reports have not all confirmed this claim and

Table 3.7 Alcoholic brain atrophy and ADH/ALDH allele
frequency (Maezawa et al. 1996)

	Cerebral atrophy	
	Minor (57)	Severe (20)
Age (y)	50.3 ± 1.0	49.7 ± 2.2^a
Total alcohol intake (kg)	$1,180 \pm 70$	992 ± 80^a
Cirrhosis (%)	42	30^b
$ALDH_2^1$	0.912	0.925
$ALDH_2^2$	0.088	0.075
ADH_2^1	0.404	0.700^c
ADH_2^2	0.596	0.300^b

Mean $\pm$ SD values.
a $p > 0.05$.
b $p < 0.05$.
c $p < 0.01$.

meta-analyses of the published data have also been contradictory (Gerlernter *et al.* 1991; Noble 1993).

The discrepancies may be related to heterogeneity of the alcohol misuse syndrome, difficulty in accurately matching patients with controls with respect to age, gender, ethnicity, clinical correlates and the possible relationship of the polymorphism to a dependent personality including polydrug misuse, criminality and antisocial and psychopathological personalities. For example, Samochowiec and colleagues (1999) found a monoamine oxidase A polymorphism in some German males with alcoholism, but this gene was subsequently related to the presence of antisocial behaviour rather than to dependence on alcohol.

So far, this brief review has solely considered the genetic basis to the development of alcoholism itself. One of the intriguing observations in alcoholics is how variable is the clinical presentation in different individuals who have seemingly drunk excessive but similar amounts of alcohol over a similar time period: there is a wide variation in organ sensitivity. Individuals may predominantly have hepatic (Lieber 1988), pancreatic (Apte *et al.* 1998), cardiovascular (Preedy and Richardson 1994), central nervous system (Charness *et al.* 1989), gastrointestinal (Bjarnason *et al.* 1984), musculoskeletal (Laitinen and Välimäki 1991) or dermatologic (Higgins and du Vivier 1994) damage with little evidence of other affected organs. It has, therefore, been postulated that, as well as a genetic component to the development of alcoholism itself, there is a genetic component to the development of specific organ damage in alcoholics. Genetic susceptibility to the development of alcohol-related cirrhosis, the Wernike–Korsakoff syndrome, psoriasis, osteoporosis, oral malignancy and cerebral atrophy have been

Table 3.8 Consequences of a genetic component to alcohol abuse

1	Understanding the aetio-pathogenesis
2	Development of prognostic and therapeutic strategies
3	Monitoring of selective organ toxicity
4	Career guidance to patients and their relatives
5	Development of pharmacotherapies
6	Gene therapy

postulated. Table 3.7 shows evidence that polymorphism of hepatic alcohol dehydrogenase (ADH) is associated with a differing prevalence of cerebral atrophy. Alcohol dehydrogenase $ADH_2{}^1$ is accompanied by lower rates of overall ethanol metabolism than $ADH_2{}^2$, and, thus, the persistently higher levels, of blood and tissue ethanol levels after a similar ethanol load in the former individuals, may be directly responsible for the increased neuronal damage in these subjects. Note that age, lifetime alcohol consumption, presence of cirrhosis or possession of the inactive acetaldehyde dehydrogenase variant are not predisposing factors. It is likely that other genetic components contributing to individual organ susceptibility will soon be identified (Ferguson and Goldberg 1997).

Consequences of a genetic component to alcohol misuse

The clear confirmation that there is a genetic component to alcoholism has several important consequences, as listed in Table 3.8. Most important from the medical care and treatment perspective is the view that alcohol misuse is purely a 'self-inflicted' disease. This change in emphasis is already having positive effects on the attitude of medical and nursing staff, public health officials and medical managers (O'Brien and McClellan 1996). It should be pointed out that environmental factors including price, availability, peer pressure, social deprivation, stress and adverse life events still have the major (approx. 75 per cent) role in the onset of alcoholism. Thus, a two-pronged approach to prevention, assessment and treatment of alcohol misuse, exploring both the nature and nurture components, remains the appropriate approach.

Chemical pathology of alcohol misuse

It should be noted that the chemical pathologist is primarily concerned with the biochemical, haematological and immunological consequences of alcohol misuse. He/she is not necessarily concerned with degree of alcohol dependence (addiction). This is important because a harmful or excessive drinker, who has none of the features of the alcohol dependency syndrome, may, nevertheless, have serious physical consequences from their excess alcohol intake. These individuals may, in the absence of a careful clinical

Table 3.9 Laboratory markers of selective organ toxicology

Organ	Marker
Central nervous system	N/A
Liver	Serum γ-glutamyl transferase, aspartate amino-transferase, carbohydrate-deficient transferrin
Pancreas and oro-gastrointestinal tract	Serum amylase, permeability and inflammatory markers
Kidney	Urinary N-acetyl-β-glucosaminidase
Skeletal muscle (acute)	Serum creatine kinase
Skeletal muscle (chronic)	N/A
Heart muscle	N/A
Skin	N/A
Skeleton	Serum procollagen peptides, urinary collagen products
Bone marrow	Erythrocyte macrocytosis, neutropaenia, thrombocytopaenia

N/A = Not available.

Table 3.10 Screening questionnaires for alcohol misuse (Cherpitel 1993)

Instrument	Sensitivity (%)	Specificity (%)
RAPS	90	75
TWEAK	87	76
AUDIT	85	82
CAGE	75	82
BMAST	29	98
Breathalyser	19	91

Instruments relate to acronyms for questionnaires listed in reference provided.

history and examination, not be recognized as drinking excessively. It is estimated that over half of patients misusing ethanol are not recognized as such by their medical attendants. The chemical pathologist has a role in assisting in the identification of such individuals.

For patients clearly recognized as suffering from alcohol misuse, the role of the chemical pathologist is to provide evidence of specific organ damage, where possible, and to monitor the response to abstinence or reduced and controlled drinking. Several of the markers, although relatively sensitive, are non-specific and, for several important organs, there is as yet no suitable blood/urine markers of damage (Table 3.9). The development of markers for these organs remains an important challenge for the chemical pathologist.

There are basically three approaches to the detection of chronic alcohol

Table 3.11 Sensitivity and specificity of laboratory tests in the detection of harmful drinking and alcohol dependency (modified from Conigrave *et al.* 1995)

Marker	Harmful drinking		Dependent alcoholism	
	Sensitivity	Specificity	Sensitivity	Specificity
Serum γ-glutamyl transferase	35	75	80	75
Aspartate aminotransferase	15	80	45	80
Carbohydrate deficient transferrin	50	95	80	95
Erythrocyte macrocytosis	25	70	45	75

misuse: (i) the use of screening questionnaires; (ii) the detection of ethanol itself in various body fluids and (iii) various markers of the consequences of chronic alcohol toxicity. Table 3.10 compares the sensitivity and specificity of various screening questionnaires with those of the ethanol breathalyser in detecting alcohol misuse by individuals attending an A&E department in the USA (Cherpitel 1993). With the exception of the brief MAST questionnaire, which is designed to identify dependent alcohol misusers, most screening instruments are highly sensitive with acceptable specificity. The determination of breath, blood or urinary alcohol levels is clearly highly specific, but, because of the relatively transitory presence of ethanol following even heavy consumption, it has an unacceptably low sensitivity.

Various biochemical tests have been developed, often following serendipitous observations, to enhance both the sensitivity and specificity of detecting alcohol misuse. Table 3.11 summarizes the current situation concerning the use of these markers. Liver toxicity tests are in routine use and are acceptably sensitive and specific in dependent alcoholics, but none of the investigations are particularly sensitive in the harmful drinking category. Carbohydrate-deficient transferrin has a high specificity and is useful in the investigations of patients with unexplained macrocytosis or abnormal liver toxicity tests, particularly if there is denial of alcohol abuse. Carbohydrate-deficient transferrin is also a valuable marker of continued alcohol consumption/abstinence in patients with associated non-alcoholic liver disease (e.g. chronic viral hepatitis, transplant rejection). It is, however, no more sensitive than other biochemical markers.

References

Apte, M. V., Haber, P. S., Norton, I. D. and Wilson, J. S. (1998) 'Alcohol and the pancreas', *Addict. Biol.*, **3**, 37–150.

Badia-Elder, N. E., Mosemiller, A. K., Elder, R. L. and Froelich, J. C. (1999) 'Naloxone retards the expression of a genetic predisposition toward alcohol drinking', *Psychopharmacology (Berlin)*, **144**, 205–12.

Bjarnason, I., Ward, K. and Peters, T. J. (1984) 'The leaky gut of alcoholism: Possible route of entry for toxic compounds', *Lancet*, **I**, 179–82.

Blum, K., Noble, E. P., Sheridan, P. J., Montgomery, A., Ritchie, T., Jagadee-swaran, P., Nogami, H., Briggs, A. H. and Cohn, J. B. (1990) 'Allelic association of human dopamine D_2 receptor gene in alcoholism', *J. Amer. Med. Assoc.*, **263**, 2055–60.

Charness, M. E., Simon, R. P. and Greenberg, D. A. (1989) 'Ethanol and the nervous system', *N. Engl. J. Med.*, **321**, 442–54.

Cherpitel, C. J. (1993) 'Alcohol and violence-related injuries: An emergency room study', *Brit. J. Addict.*, **88**, 79–88.

Chick, J. (1996) 'Medication in the treatment of alcohol dependence', *Advances Psychiat. Treat.*, **2**, 249–57.

Cloninger, C. R. (1987) 'Neurogenetic adaptive mechanisms in alcoholism', *Science*, **236**, 410–16.

Conigrave, K. M., Saunders, J. B. and Whitfield, J. B. (1995) 'Diagnostic tests for alcohol consumption', *Alcohol*, **30**, 13–26.

Cotton, N. S. (1979) 'The familial incidence of alcoholism: A review', *J. Stud. Alcohol*, **40**, 89–116.

Day, C. P. and Bassendine, M.F. (1992) 'Genetic predisposition to alcoholic liver disease', *Gut*, **33**, 1444–7.

Ferguson, R. A. and Goldberg, D. M. (1997) Genetic markers of alcohol abuse. *Clin. Chim. Acta*, **257**, 199–250.

Gerlernter, J., O'Malley, S., Risch, N., Kranzier, H. R., Krystal, J., Merikangas, K., Kennedy, J. L. and Kidd, K. K. (1991) 'No association between an allele at the D2 dopamine receptor gene (DRD2) and alcoholism', *J. Amer. Med. Assoc.*, **266**, 1801–7.

Goodwin, D. W. (1981) 'Genetic component of alcoholism', *Ann. Rev. Med.*, **32**, 93–9.

Higgins, E. M. and du Vivier, A. W. P. (1994) 'Cutaneous disease and alcohol misuse', *Brit. Med. Bull.*, **50**, 85–98.

Higuchi, S., Matsushita, S., Muvamatsu, T. and Hayashida, M. (1996) 'Alcohol and aldehyde genotypes and drinking behaviour in Japanese', *Alcohol Clin. Exp. Res.*, **20**, 493–7.

Laitinen, K. and Välimäki, M. (1991) 'Alcohol and bone', *Calcif. Tiss. Int.*, **49**, 570–3.

Li, T-K. and McBride, W. J. (1995) 'Pharmacogenetic models of alcoholism', *Clin. Neurosci.*, **3**, 182–8.

Lieber, C. S. (1988) 'Biochemical and molecular basis of alcohol-induced injury to liver tissues', *N. Engl. J. Med.*, **319**, 1639–50.

Maezawa, Y., Yamauchi, I. M., Seashi, Y., Mizuhara, Y., Kimura, T., Toda, G., Suzuki, I. H. and Sakurai, S. (1996) 'Association of restriction fragment poly-morphism in the alcohol dehdyrogenase 2 gene with alcoholic brain atrophy', *Alcohol Clin. Exp. Res.*, **20** (Suppl.), 29–32A.

Noble, E. P. (1993) 'The dopamine receptor gene: A review of association studies in alcoholism', *Behav. Genet.*, **23**, 119–229.

O'Brien, C. P. and McLellan, A. T. (1996) 'Myths about the treatment of addiction', *Lancet*, **347**, 237–40.

Plomin, R., Owen, M. J. and McGuffin, P. (1994) 'The genetic basis of complex human behaviours', *Science*, **264**, 1733–9.

Preedy, V. R. and Richardson, P. J. (1994) 'Ethanol-induced cardiovascular disease', *Brit. Med. Bull.*, **50**, 152–63.

Samochowiec, J., Lesch, K. P., Rottmann, M., Smolka, M., Syagailo, Y. V., Okladovna, O., Rommelspacher, H., Winterer, G., Schmidt, L. G. and Sander, T. (1999) 'Association of a regulatory polymorphism in the promoter region of the monoamine oxidase A gene with antisocial alcoholism', *Psychiat. Res.*, **86**, 67–72.

Shibuya, A. and Yoshida, A. (1988) 'Genotypes of alcohol-metabolizing enzymes in Japanese with alcohol liver diseases: a strong association of the usual Caucasian-type aldehyde dehydrogenase gene (ALDH1(2)) with the disease', *Am. J. Hum. Genet.*, **43**, 744–8.

Whitfield, J. B. (1997) 'Acute reactions to alcohol', *Addict. Biol.*, **2**, 377–86.

Williams, J. T., Begleiter, H., Porjesz, B., Edenberg, H. J., Foroud, T., Reich, T., Goate, A., Van Eerdewegh, P., Almasy, L. and Blangero, J. (1999) 'Joint multipoint linkage analysis of multivariate qualitative and quantitative traits. II. Alcoholism and event-related potentials', *Amer. J. Human Genet.*, **65**, 1148–60.

Wilson, W. (1999) 'Geroff my tail!', *New Scientist*, **163**(31 July), 46–7.

Yoshida, A., Dave, V., Ward, R. J. and Peters, T. J. (1989) 'Cytosolic aldehyde dehydrogenase (ALDH$_1$) variants found in alcohol flushers', *Ann. Hum. Genet.*, **53**, 1–7.

Chapter 4

Alcohol and the brain

William P. Watson and Hilary J. Little

Key points

- *Alcohol affects many different neurotransmitter systems.*
- *This helps to explain the many differing actions of alcohol.*
- *Alcohol decreases the effects of excitatory amino acid transmitters.*
- *Alcohol increases the actions of the inhibitory transmitter gamma-aminobutyric acid (GABA) (γ-aminobutyric acid) in some neurons. Alcohol also affects transmission involving enkephalins, dopamine systems and 5-HT pathways.*
- *Alcohol decreases perceived anxiety and stress.*
- *At high doses alcohol is a general anaesthetic.*
- *Prolonged intake of alcohol causes tolerance and dependence.*
- *On cessation of long-term drinking a withdrawal syndrome is seen: this can involve anxiety, tremors, convulsions and hallucinations.*
- *Many people drink alcohol but do not become dependent on it.*
- *The mechanism of tolerance to alcohol involves many transmitter systems as well as environmental cues.*

The pharmacological properties of alcohol

Ethanol has a variety of actions on the central nervous system (Figure 4.1). These are produced at concentrations considerably higher than those at which most psychoactive drugs are effective. The majority of drugs act at pM, nM or μM concentrations (i.e. 10^{-12}, 10^{-9}, 10^{-6} molar), but ethanol acts at mM concentrations (i.e. 10^{-3} molar or upwards). Figure 4.1 illustrates the concentrations recorded in the blood at which alcohol produced a range of effects on the brain. These concentrations are given both in molar units and in mg/100 ml, the units used in defining legal limits for driving.

From as long ago as the turn of the twentieth century, it was realized the initial target site in the body for alcohol might differ from that of other drugs (Meyer 1899). There do not appear to be alcohol-specific protein receptors to which it can bind. This type of action is described as 'non-specific' – in this

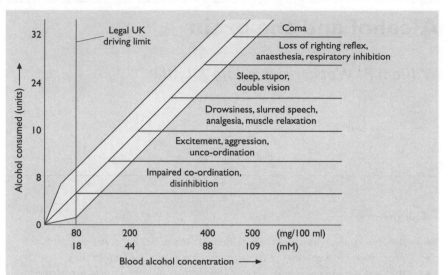

Figure 4.1 A graph relating amount of alcohol consumed with blood ethanol concentrations, indicating physical symptoms at each blood level. Three lines are shown to represent the variation in the distribution of ethanol in the body.

context this term is used in a precise pharmacological sense to indicate actions that are not produced through protein receptor sites, and it does not mean that alcohol has actions at all sites in the brain. In fact, the effects of alcohol on system transmission in the central nervous system are in many instances very selective.

Even now, the initial target site of alcohol is the subject of much controversy. It was originally suggested in 1899 that alcohol acted on lipids in cells. The site of action may be the lipids that make up the neuronal membranes, and it has been suggested that alcohol may alter the fluidity or amount of molecular movement of the lipid molecules. Recent theories, however, have suggested that it may act on proteins in the membranes that are involved in synaptic transmission. It is also possible that these effects are produced via actions on the membrane lipids, particularly those that make up the immediate environment surrounding the protein molecules. Alternatively, the site of action of alcohol may be hydrophobic 'pockets' within the proteins in the neuronal membrane.

Alcohol affects many different transmitter systems within the central nervous system. In this, it is similar to general anaesthetic drugs (e.g. nitrous oxide [laughing gas] and ether). Both of these anaesthetic gases were used during the nineteenth century at parties for their euphoriant effects.

Alcohol can be described as a stimulant or as a depressant; neither description is comprehensive, as it has a range of different effects that can vary according to the circumstances. At low blood concentrations, it can have excitant effects. These are sometimes known as 'disinhibitory', as people become less concerned with the various anxieties that normally inhibit behaviour. As the concentrations rise, alcohol has an overall depressant effect on the central nervous system. The widely used phrase 'loss of inhibition' does not refer to the effects on central synaptic transmission; inhibitory transmission is actually increased by alcohol. However, alcohol has euphoriant and anxiolytic properties that may result in relaxation of the normal constraints on behaviour.

Why do people drink alcohol?

Alcohol is consumed socially by a large percentage of the UK population. A proportion of them become dependent on the drug, and the mechanisms that may be involved in this progression are discussed below. However, a large majority of the population consume alcohol regularly, in small or large amounts, without becoming dependent. The reasons why people begin to drink alcohol (initiation) should be distinguished, from a mechanistic point of view, from the reasons alcoholics continue to drink alcohol, despite its adverse effects, and these two aspects will be discussed separately.

Among the reasons that people give for drinking alcohol socially are its anxiolytic actions and its euphoric effects. Alcohol decreases perceived anxiety, particularly in social situations. This is known as the 'tension-reduction' theory, which suggests that alcohol is used as a form of self-medication. Alcohol, like many drugs of addiction including psycho-stimulants and opiates, can cause euphoria, the pleasant feeling experienced at certain levels of intoxication. The effects of alcohol, especially on mood, are affected by the environment in which it is consumed; for example, it can be less euphoric in a laboratory setting, or when drunk alone, than when consumed in a social situation.

Alcohol also has properties which are described as positively 'reinforcing' The term 'reinforcement' is used to describe the effect of a drug, or situation, that results in repetition of the behaviour that preceded the experience. How far this is related to the other properties of alcohol is uncertain. The reinforcing effect can involve the feeling we know as 'pleasure', but this is not necessarily the case; for example, some drugs which stimulate the brain, but which are not thought to cause euphoria, have been shown to have reinforcing properties in laboratory animals. Alcohol can also be aversive; for example, initially, most people and most animals dislike the taste.

Other factors contribute to alcohol consumption (Figure 4.2), particularly in adolescents, including social influences such as peer pressure and sensation seeking. There are large individual differences in the tendency to indulge in a potentially dangerous behaviour, which we may now be beginning to understand from a biochemical point of view.

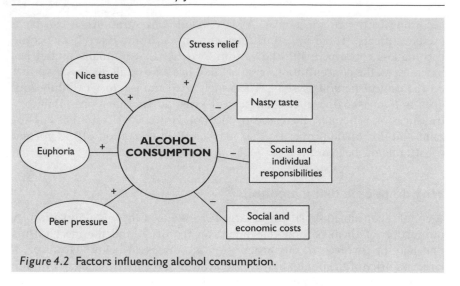

Figure 4.2 Factors influencing alcohol consumption.

Pharmacological effects of alcohol on the brain

Many, though not all, of the effects of alcohol on the brain can now be explained in terms of its effects on central synaptic transmission. The acute effects of alcohol, familiar to most people, are mood changes, loss of motor co-ordination and amnesia. These can be pleasant or unpleasant depending on the circumstances and the amount drunk. The gastric irritation experienced is due to local effects on the digestive system. A 'hangover' is due to a combination of this gastric irritation, to dehydration and to the unpleasant effects of the major metabolite of alcohol, acetaldehyde.

In psychological terms, alcohol has dual actions, being both rewarding and aversive. Its effects at any one time depend on the situation, the state of mind of the person consuming it and the previous experience of that person with alcohol. The many actions of alcohol on the brain are due to its diverse effects on synaptic transmission involving a variety of neurotransmitters in the central nervous system, illustrated in Figure 4.3.

Alcohol decreases the effects of excitatory amino acid transmitters and increases the actions of the inhibitory transmitter GABA. Both these effects are likely to play a role in its anxiolytic properties.

The potentiation of GABA transmission is seen at low concentrations, but is not apparent on all the neurones which release GABA as a transmitter, nor in all areas of the brain. In its effect of potentiating the inhibitory transmitter GABA, alcohol resembles the benzodiazepine drugs, but the effects of alcohol are far more complex than those of benzodiazepines and involve many other systems.

Alcohol decreases the effects of excitatory transmitters, glutamate and aspartate, which mediate excitatory transmission in practically all areas of

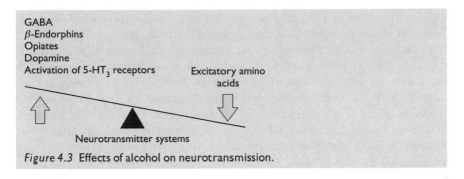

GABA
β-Endorphins
Opiates
Dopamine
Activation of 5-HT₃ receptors   Excitatory amino
                                                    acids

Neurotransmitter systems

Figure 4.3 Effects of alcohol on neurotransmission.

the brain. It has particular effects on the synapses that use the NMDA receptor subtype, and higher concentrations are needed to block transmission via the AMPA or kainate subtypes of receptor. The NMDA-receptor blocking actions may be relevant to the amnesic properties of alcohol and, possibly, the anxiolytic actions.

Alcohol also alters the synaptic transmission systems that involve endogenous opiates, called enkephalins and endorphins. It was demonstrated, in 1975 in Aberdeen, that the chemical compounds which these systems use as neurotransmitters have similar actions to those of morphine and heroin. Activation of endogenous opiate transmission by alcohol has been found to occur at relatively low concentrations and is thought to be involved in its euphoric effects. This may be the basis of the effects of the opiate antagonist naltrexone in decreasing alcohol drinking in some alcoholics. Alcohol also causes release of β-endorphin from the hypothalamus and the pituitary gland, but the significance of this action is uncertain.

Alcohol has many effects on monoamine transmitters, in particular activating the dopamine and 5-HT (serotonin) pathways. The mesolimbic dopamine system in the brain is thought to be involved generally in 'reward' processes and specifically in the reinforcing effects of drugs, although its exact role is uncertain. Alcohol is known to have activating effects on neurones in the ventral tegmental area, where the cell bodies of the mesolimbic dopamine system are located, and it increases extracellular dopamine concentrations in the nucleus accumbens, one of the main projection areas of the ventral tegmentum in the laboratory rat. This effect of alcohol in increasing the firing rate of dopaminergic neurones, particularly in the ventral tegmental area, may be the basis of the reinforcing action of alcohol.

Alcohol causes many alterations in 5-HT transmission, including potentiation of the activation of 5-HT₃ receptors. Changes in the many central systems which contain the plethora of different subtypes of receptors for 5-HT may be involved in the mood changes commonly seen after alcohol. Drugs which potentiate 5-HT transmission by blocking the re-uptake process (SSRIs, selective serotonin re-uptake inhibitors) have been shown in the laboratory to decrease alcohol consumption, as have drugs with actions at certain of the receptor subtypes, but clinical trials of these drugs have not so far demonstrated clear beneficial effects in alcoholics.

Potentially, many other effects are caused by alcohol on different transmitter systems, but the important ones are those, described above, that are produced at blood concentrations found in the body when alcohol is consumed. At high doses, alcohol is a general anaesthetic and was used for many years for this purpose before the anaesthetic gases were introduced. However, it has dangerous effects on the respiratory and cardiovascular systems at doses just above the threshold for anaesthesia; so, alcohol is no longer used for this purpose.

Neuronal effects of prolonged consumption of alcohol

Current theories of dependence on alcohol suggest that adaptive changes take place in central neurones, including the mesolimbic dopamine system, which affect the brain's response to alcohol, but do these effects of alcohol result behaviourally in further drinking? With prolonged alcohol consumption, its actions are modified at a cellular level by adaptive responses in central neurones. At the level of the individual, the alcohol withdrawal syndrome and tolerance to alcohol are characteristic of alcohol dependence. It has been known for some time that neuronal adaptations develop in parallel to tolerance to the behavioural effects of alcohol and to the withdrawal syndrome, but the importance of these two types of phenomena with regard to alcohol dependence is now the subject of controversy. In recent years, research has been concentrated on understanding the neurophysiological mechanisms that underlie dependence on alcohol. Figure 4.4 illustrates diagrammatically the processes of adaptation to alcohol and the withdrawal syndrome.

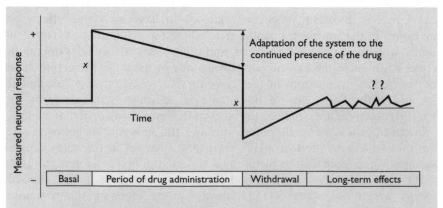

Figure 4.4 The process of neuronal adaptation to the continued presence of alcohol and its subsequent removal.

Alcohol tolerance

On long-term alcohol intake, tolerance occurs as the brain becomes 'accustomed' to the presence of the alcohol. This means that physiological changes take place to minimize the effects of alcohol – 'homeostasis' – just as when adaptation occurs to temperature or low oxygen levels. Tolerance to alcohol is not just a single adaptive process, but occurs in several forms. The first distinction is between 'metabolic' and 'functional' tolerance. Metabolic (sometimes known as distributional tolerance) is due to changes in the metabolism or distribution of a drug, resulting in decreases in the concentrations reaching the site of action. Functional tolerance (also known as 'cellular' or 'pharmacodynamic' tolerance) describes changes at the site(s) of action of a drug, so that the effects of a particular concentration are reduced. A proportion of tolerance to alcohol is metabolic, due to changes in enzyme action, but alcoholics and heavy drinkers can function after doses of alcohol that would render most people unconscious. Although many adaptive changes in neurones affected by alcohol have been described, it is not clear which of these are responsible for functional tolerance.

There is also an environmental influence on drug tolerance, which is seen with alcohol (Goudie and Griffiths 1986). Drug intake in surroundings in which a person is used to taking a drug results in greater tolerance than if the same dose of drug is consumed in novel surroundings. This effect is thought to contribute to fatalities when illegal drugs are taken in an unfamiliar setting. While the context of drinking can contribute to alcohol tolerance, the cellular mechanisms involved are not understood. Another component of tolerance is produced by performance of a task while under the influence of alcohol. This can be demonstrated experimentally by performance of motor tasks after alcohol consumption.

Certain peptides have been found experimentally to influence tolerance to alcohol, although the mechanism of these effects is not yet fully understood. Administration of oxytocin, while alcohol is still being given, decreases alcohol tolerance. Vasopressin, when given after cessation of alcohol treatment, prolongs the duration of the tolerance, which normally dissipates within a few days.

The alcohol withdrawal syndrome

The withdrawal syndrome is produced when drinking is stopped after prolonged high consumption. Seizures can occur, which may even be fatal if not treated, but pharmacological intervention, normally with benzodiazepines, can prevent such seizures and many of the other manifestations of the withdrawal syndrome. However, such treatment does not prevent a return to abnormal drinking, and the significance of the withdrawal syndrome in maintaining dependence is unclear.

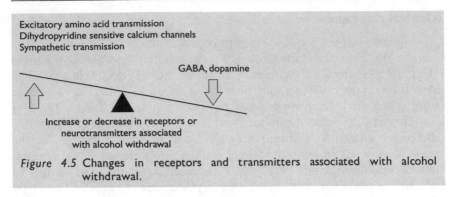

Figure 4.5 Changes in receptors and transmitters associated with alcohol withdrawal.

Research over the last decade has elucidated the neuronal changes that underlie the alcohol withdrawal syndrome. In has been found that excitatory amino acid transmission is increased, with increased excitatory postsynaptic potentials which involve receptors of the NMDA and the AMPA/kainate subtypes. The density of receptors of the NMDA subtype is also increased. Increased activity of voltage-sensitive calcium channels has also been demonstrated, particularly the long-duration, high-voltage activated channels that are sensitive to blockade by dihydropyridine compounds. These compounds have little effect on neurones in the central nervous system in normal circumstances, but the calcium channels on which they act are now thought to be important in the adaptive changes in the brain to exogenous disturbance, such as that produced by drugs. Decreases in inhibitory transmission, however, do not appear to be as important in the genesis of the withdrawal syndrome as was originally thought.

The alcohol withdrawal syndrome is usually controlled therapeutically with benzodiazepine drugs, chlormethiazole or barbiturate compounds (Figure 4.5). These prevent the seizures and may be considered to be substituting for alcohol. Although such treatment can reduce withdrawal symptoms, it is not always effective and the drugs used carry a risk of serious interactions with alcohol if drinking is resumed, as well as the potential for producing dependence themselves. New approaches are being explored. The value of calcium channel antagonists is under investigation, although the compounds of this type currently available are not selective for the central nervous system. Other novel types of compound, such as gabapentin, with more selective actions, have been demonstrated in the laboratory to be effective against both the convulsive and anxiety-producing elements of the syndrome and may have advantages in the treatment of the acute withdrawal phase.

Prolonged effects of alcohol

A predominant characteristic of dependence on drugs is that addicts can stop using the drug, recover from the acute withdrawal phase (detoxification), and may go for weeks or months of abstinence, but then relapse back into drug taking. This suggests that the acute withdrawal phase, and the neuronal changes that underlie the withdrawal syndrome, are not those responsible for relapse drug taking. Research is, therefore, beginning to concentrate on more prolonged effects of chronic alcohol consumption.

Release of dopamine in the nucleus accumbens is decreased during alcohol withdrawal in rats, and this has been suggested to be responsible for the dysphoric effects of withdrawal. However, measurement of dopamine concentrations over time has shown that the decrease in release appears to outlast the period over which the behavioural signs of acute withdrawal are seen. Decreases in the firing rate of ventral tegmental neurones also persist long after the cessation of alcohol withdrawal signs in rodents.

There has been much interest recently in 'sensitization' to drugs of abuse, the increase in sensitivity when drugs, such as amphetamine, are taken repeatedly. This phenomenon is thought to be of importance because it is a very long-term, possibly permanent change (Robinson and Berridge 1993), but its clinical relevance is uncertain. Alcohol has been reported to cause similar changes to those seen after prolonged intake of other drugs of abuse, but it is of interest to note, in view of the slow onset of alcohol dependence, that such changes were seen only after very long periods of alcohol intake.

Individual differences in the effects of alcohol

Why does alcohol cause alcohol dependence? This question should really be phrased 'why does alcohol cause alcohol dependence *in some people*?' The crucial question is why only a small minority of people who drink alcohol become dependent on it. Three main factors are thought to be involved in this individual variation: genetic influences, the effects of stress and the influence of the environment.

Genetic aspects of alcoholism

The genetic aspects of alcoholism are described in full in Chapter 3 and so will not be detailed here. This aspect is under intensive research and the neuronal effects of alcohol in rodent strains that differ widely in their willingness to consume alcohol have provided models for genetic studies.

Involvement of stress

Stress and adverse life events are now thought to play an important role in the development of alcohol dependence. As there are many forms of stress, and many factors that contribute to the development of drug dependence, this relationship has been difficult to examine in humans, but several studies have shown higher incidence of adverse 'life events' in alcoholics compared with age-matched controls. Stress is known to be one of the factors that can precipitate relapse in abstinent alcoholics.

Recent experimental evidence has shown that the adrenocorticoid hormone, corticosterone (the rodent equivalent of cortisol), has effects on alcohol consumption in rodents. High blood levels of corticosterone are associated with high alcohol consumption and drugs that decrease corticosterone levels decrease voluntary consumption of alcohol. There have been reports of defects in control of adrenocortical hormone release in alcoholics, but the basis of these is not known. Some alcoholics have high basal circulating levels of cortisol, but lower concentrations in response to stressful situations.

There is a high incidence of increased cortisol levels in depression, and in tests a lack of effect of dexamethasone (a synthetic analogue of glucocorticoid hormone) in suppressing release of cortisol has been used in diagnosis of depression. There is some degree of co-morbidity of depression and alcoholism, but it is not clear whether this is due to pre-existing depression or is the result of excessive alcohol consumption.

A group of rats or mice placed together will rapidly assume a social rank order. It has been found that subordinate animals have much higher voluntary alcohol consumption than dominant animals. Subordinate animals have high basal plasma levels of corticosterone but lower levels in response to stress. Observations such as these are providing a foundation for the investigation of the mechanisms by which stressful experiences may contribute to loss of control of drinking and research in this area may prove extremely valuable in our understanding of this area.

Environmental influences

It is now recognized that the effects of a drug are affected by many factors, and are not just fixed. These factors include prior experience with alcohol and the influence of the environment in which the drug is taken. A return to the surroundings in which an abstinent alcoholic is used to drinking is very likely to produce a relapse. Our understanding of the neuronal mechanisms underlying conditioning to environmental cues is at an early stage, but may in future contribute much to our understanding of relapse drinking.

Commonalities between drugs of abuse

The initial target sites within the central nervous system differ for the various types of drugs of abuse; for example, opiate drugs act on opiate receptors, nicotine on nicotinic receptors and alcohol on the many sites within neuronal systems described above. However, a considerable amount of recent evidence (summarized by Iversen 1999) has suggested

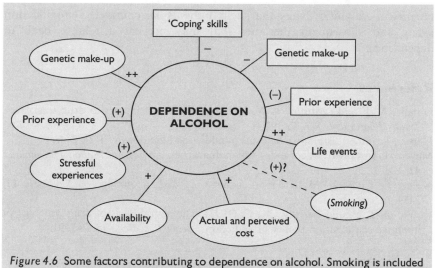

Figure 4.6 Some factors contributing to dependence on alcohol. Smoking is included as an example of a phenomenon closely linked to alcohol dependence, but not thought to contribute to its development.

that there may be common mechanisms in the neuronal basis of dependence on different types of drugs. The mesolimbic dopamine system has been suggested to be of major importance in such dependence.

Many current theories suggest that drugs of dependence cause changes at the level of the mesolimbic dopamine system, which is thought to be involved in the 'rewarding' effects of natural stimuli, such as food. These theories suggest that conditioned changes in this system result in craving for the drugs. However, as described above, the mesolimbic system is also activated by other sensory input, including stressful experiences, and it is likely that the basis for dependence is not as simple as the original theories suggested.

Other evidence suggesting commonalities among different types of drugs of dependence in their actions on the brain include similar neurochemical changes reported at the level of individual neurones, after prolonged intake of the drugs, and common neuronal patterns of sensitization (Nestler 1992).

Conclusion

The pharmacology of substance use can help to explain why different individuals respond to one substance in different ways and why one person responds differently to that substance at different times. For people who have become dependent on that substance, pharmacological research may identify therapeutic drugs that help patients to complete detoxification safely and subsequent therapies that may prevent a relapse back to dependence.

References

Goudie, A. J. and Griffiths, J. W. (1986) 'Behavioural factors in drug tolerance', *Trends Pharmacol. Sci.*, **7**, 192–6.

Iversen, L. (1999) 'Brain mechanisms in addiction', *EuroBrain*, **1**(2), 2–4.

Meyer, H. H. (1899) 'Theorie der Alkoholnarkose', *Arch. Exp. Pathol. Pharmakol.*, **42**, 109–18.

Nestler E. J. (1992) 'Molecular mechanisms of drug addiction', *J. Neurosci.*, **12**, 2439–50.

Robinson, T. E. and Berridge, K. C. (1993) 'The neural basis of drug craving: An incentive-sensitization theory of addiction', *Brain Res. Rev.*, **18**, 247–91.

Chapter 5

Alcohol on the mind

Woody Caan

Key points

An understanding of alcohol's effects involves all the following elements:

- *Psychopharmacology (mediated by a variety of receptor sites in the brain).*
- *Individual differences (like personality).*
- *Cognitive changes (learning about drinking).*
- *Social psychology (the role of drinking in groups).*
- *Alcohol history (the cumulative impact of drinking over time).*

The dose of alcohol, the context within which someone is drinking and the duration of alcohol use are therefore important considerations for understanding anyone who is intoxicated. For a significant minority of humankind, alcohol use eventually forms new ways of thinking and behaving which appear to be surprising and 'pathological'. However, most people with drink problems show distinct patterns of psychological change, which can illuminate their condition, drunk or sober.

Why does it help to understand the psychology?

Most readers will have at least some personal experience of the psychological effects of ethyl alcohol. Quite a few readers will have experienced problems during alcohol intoxication and some distress afterwards, and yet, for over 90 per cent of the British population, drinking is a regular element of their behaviour. Overwhelmingly, pleasurable times like parties and celebrations are associated with alcohol use, and yet so are nightmarish events like domestic violence or the 'hit and run' driver who crushes a child while drunk. This chapter will provide one framework into which both the familiar and the puzzling effects of alcohol on the mind will fit.

Alcohol use seen 'from the outside': effects on synapses, systems and single behaviours

At high doses (peak blood concentrations of ethanol over about 200 mg/ 100 ml), drinking alcohol always acts as a central nervous system depressant which reduces overall brain excitation: eventually, an anaesthetic state can be reached which is followed at only slightly higher doses (about 400 to 500 mg/100 ml) by death when the brainstem systems supporting normal breathing, blood pressure and body temperature cease to maintain life. For a few unfortunate children who overdose accidentally and for the thousands of adults who commit suicide with the aid of alcohol (usually in combination with other central nervous system depressants like sleeping pills), this is the end of the story. However, at the lower doses most of us employ (after, say, one to two glasses of wine, blood ethanol concentrations reach 15–30 mg/100 ml), much subtler psychotropic effects are present. Most people in most settings appear more relaxed and confident when they begin drinking. Alcohol acts at many different sites in the brain (Samson and Harris 1992), which is why attempts to produce a single 'amethystic' drug to sober up an intoxicated person have all failed. However, let us start by considering one receptor site which has been especially well characterized.

Inhibitory synapses in many parts of the brain release the neurotransmitter GABA, and alcohol binds to the GABA-A receptor in a way which increases that inhibition of the target, post-synaptic neurones. Many drugs that reduce anxiety (benzodiazepines, barbiturates, valerian, zopiclone) share with alcohol this modulation of GABA actions, although each class of 'anxiolytic' drug binds to the GABA-A receptor in a slightly different fashion.

Increasing the effects of GABA has three actions on animal behaviour which can be observed reliably: we see progressively disorganized *movement*, *sleepiness* and in 'conflict' situations, which combine the possibility of receiving *either* reward *or* punishment, there appears an increasing tendency to *try for the reward* and *risk the punishment*. GABA neurones are especially common in the cerebellum and many of the motor effects of alcohol seen in humans (unsteady balance, staggering/weaving gait, sweeping unco-ordinated gestures and slurred speech) are also typical of cerebellar dysfunction. Such poor co-ordination can appear at very low doses of ethanol, which relates to the unrecognized hazards of drink driving and operating machinery at work even when the drinker feels 'unaffected' by alcohol at an emotional level. Attention also suffers, with most people becoming less alert when they drink; this hypnotic action is useful when drinking a 'nightcap' before bedtime, but is highly undesirable in the pilot of a jumbo jet. Although it has been claimed that alcohol has good analgesic actions in children (e.g. in infants undergoing circumcision rituals), it takes very high doses in adults to reduce the sensation of pain. What low doses of alcohol can do very effectively is to shift the perceived balance between pleasure and pain (either physical or emotional distress). In the initial stages of intoxication, people may appear disinhibited (talking or laughing more than usual, sharing grandiose 'bar room philosophy' with strangers) and socially shy individuals may appear more relaxed than usual entering a crowd, but these obvious and transient behaviour changes may be less interesting than the subtle but consistent change in the way people weigh up the 'costs and benefits' in a conflict situation where they feel torn between different courses of action.

An imaginary tale of teenage angst ...

One evening, an awkward 14-year-old boy Ned has been invited to his older friend Dick's house for the first time, for a party. Soon after Ned arrives, Dick disappears leaving him alone in a mixed sex group of strangers, who are beginning to dance enthusiastically. Ned feels attracted to a group of miniskirted girls who are hanging around the edge of the dance floor, giggling, and longs to ask the least spotty one for a dance, but he fears she will reject his clumsy advance dismissively followed by all the girls laughing at him. A safer alternative seems to be a retreat to the kitchen where a solitary boy called Woody is wearing a Star Trek tee shirt and might be willing to discuss Klingons for the rest of the evening, although this seems an unexciting type of party. Fortunately, the sideboard offers many cans of strong lager, and after consuming two cans rapidly (about 40 grams of ethanol) Ned not only feels a lot more cool, but both the girls and he seem to have more sex appeal and humiliation seems only a remote possibility. He approaches the first girl with 'Wanna dance?' ...

In a series of studies over the last decade, the Health Education Authority has shown that most British children establish their pattern of substance use (or non-use) before the age of 16 and the recent report of Alcohol Concern (1997) shows that 17 per cent of 11- to 15-year-olds are drinking 'regularly' about 60–70 g of ethanol each week (3.5 pints of beer). Early adolescence is a time when many of our behaviours are tested and practised, often in situations of uncertainty and anxiety, and, if alcohol becomes associated with reassuring or gratifying experiences at this critical stage of development, it is not surprising that drinking behaviour persists in most adults. Whatever their regular level of alcohol consumption, 90 per cent of all teenagers drink 'to celebrate', but valuing certain effects can predict the highest dose users by the time they are 16 to 19 years old; for example, the girls who use the most alcohol are more than three times as likely to drink 'to be confident' as the girls using lesser amounts (Foxcroft and Lowe 1993). By the time they reach college age, 50 per cent of Scottish medical, nursing, education and psychology students are already exceeding the Department of Health's 'sensible limits' for drinking (i.e. over 21 unit drinks per week for men and over 14 units for women; Engs and Van Teijlingen 1997). Of course, as we have seen above, our very perceptions of risk evaporate with drinking.

With drinking, behaviours which are not typical of sober states because of their high physical and emotional cost may emerge (e.g. alcohol use often precedes exposure to HIV; Caan 1992). At this level of functioning, belated feelings of regret when sober are irrelevant to the behaviours observed; for example, in one study we did on male medical students, every one recalled committing acts like vandalism or theft when bingeing on alcohol, and nearly all had estranged a girlfriend after a quarrel when intoxicated, but in not a single case did this alter their pattern of drinking. At a national level, the British Crime Survey found that in 44 per cent of all violent incidents the assailant is 'drunk' (Alert 1994). At a public health level, personal safety is affected by drinking in many ways (Holder 1997) including the way a child's perception of a safe home environment is undermined by family disruption related to alcohol. It is not surprising that Social Services have to tackle problems related to drinking in every age of client (Platt 1997); for example, Platt detailed one community service dealing with 30 alcohol cases a week which saw almost as many *relatives* 'affected by alcohol abuse' as they saw clients who had 'alcohol problems' themselves.

The picture of alcohol effects above would not be very different if other GABA receptor modulators like the benzodiazepines had been substituted for ethanol. However, we know that alcohol use alters many microscopic systems in the brain, in particular repeated use changes voltage-controlled channels (e.g. the L-calcium channel), receptor-mediated channels (e.g the NMDA activated calcium channel) and the synthesis of neurotransmitters (e.g. 5-HT and enkephalins). The subjective effects of alcohol described below are almost certainly the result of a complex *profile* of actions across many interrelated systems (Altman *et al.* 1996). It is at this next level that *individual differences* like personality and upbringing are most apparent in colouring the effects of alcohol on the mind.

In approaching the interaction of alcohol and mind, below, I will not attempt here to promote any particular *theory of mind*, but rather leave the reader with the stance of the Indian hero Arjuna (1995 translation):

Only thy Spirit knows thy Spirit: only thou knowest thyself.

Alcohol use seen 'from the inside': effects on mood, memory and motivation

A simple and popular explanation of taking alcohol and other sought-after substances is that the experience of using them is inherently pleasurable; in the quotable phrase of Trevor Robbins 'Drugs hijack the natural brain reward systems that control behaviour' (Nash 1997). Associated with this theory are simple models of pharmacology on the lines of: euphoria = increased dopamine release; alcohol increases dopamine; more dopamine = more pleasure.

Alcohol does initially raise the mood of some people, but there are enormous variations in the subjective effects between different individuals, and at different times in the same individual; for example, depending on the setting in which they are drinking (say, a rugby match or a funeral) and their expectations (say, optimism or dread). The same blood alcohol level may be associated with a high mood at the beginning of a drinking episode and a low mood as that episode ends (Goodwin 1994). The reward/dopamine hypothesis is sometimes elaborated by suggesting in addition that alcohol initially boosts brain 5-HT activity and then blocks the NMDA receptors, and that the pattern over time of these three systems may determine not only what is perceived as pleasurable, but also what we *remember* later about the pleasure we have had. With our present knowledge of psychopharmacology, this is about as far as neuroscience can take us in understanding alcohol, and it is based on a lot of extrapolation from the rat's brain to the human mind. In terms of understanding common drink problems in humans, it is important to note that, contrary to the simple reward theories, the people who are *most* at risk of developing problems are those who experience the *least* subjective 'high' when they try alcohol (Schuckit 1994) and, in the population of those who will need treatment for alcoholism, their brains have *fewer* receptors sensitive to dopamine than other people (Volkow *et al.* 1996).

Four different approaches have taken our understanding of alcohol use much further:

- large-scale studies of personality and the different attributions given to alcohol;
- large-scale studies of genetics and the patterns of motivation to drink;
- individual life stories and the common themes that emerge about life events;
- cognitive studies of learning and unlearning aspects of individual's alcohol use.

The personality with which we enter adolescence affects the significance we attribute to drinking and over time drinking can affect the maturation of our personality. The work of Marc Schuckit and colleagues (1995) using the Tridimensional Personality Questionnaire has been highly informative about individual features associated over time with harmful use of alcohol. A personality *Type A* characterized by 'reward dependence' has a later onset of alcoholism than the *Type B* personality characterized by 'harm avoidance' The people with an earlier onset of drink problems (often starting hazardous drinking before 16 years) tend to develop a more severe dependence on alcohol as adults, to use a variety of substances and to have generally poorer global functioning in adulthood. The concepts of reward dependence and harm avoidance build on the pioneering work of Cloninger (1987), who described *Type I* (adult onset) and *Type II* (juvenile onset) personalities at risk of developing alcoholism. However, Cloninger's Type II group included a number of young people with a history of anti-social behaviour (like fighting when drunk), but Schuckit's description of his Type B excludes this minority, well known to police and probation services, giving these 'juvenile delinquents' a separate category: at-risk drinkers with an Anti-Social Personality Disorder.

Comparing a forensic sample of young people arrested for drink-driving offences with three control groups watching videotapes of driving revealed consistent differences in the riskiness they attributed to hazards when driving (Deery and Love 1996). When sober, the young offenders perceived risks arising from other motorists in a similar way to the controls, but the offenders attributed much less risk related to their own hazardous driving than the controls attributed to their equivalent behaviours. With blood alcohol levels of 50 mg/100 ml, the drink drivers not only continued to discount the hazards over which they had active control, now they also attributed little risk to the behaviour of other motorists. In other words, these young offenders (who had all experienced accidents while driving!) actually felt safer all round when they were driving while intoxicated. The relationship in young people of harm avoidance, alcohol and deliberate self-harm is of major practical concern to many agencies today, from the Samaritans to the Health Education Authority. A large cohort of people who had shown any deliberate self-harm was followed up for 18 years by De Moore and Robertson (1996), during which time fifteen suicides occurred. The biggest component of this group who eventually committed suicide had first presented to doctors as teenage males under the influence of alcohol. By the time they died:

> substance use appeared to be a way of life,
> or a means of dealing with or avoiding conflict.
>
> (De Moore and Robertson 1996)

Suominen and colleagues (1997) studied people with or without any psychiatric diagnosis who had attempted suicide. Suicide attempters with major depression acted with much more suicidal intent than the attempters with alcoholism, but alcoholic attempters were much more likely to show 'impulsive' personalities than depressed attempters (68 per cent impulsive vs 24 per cent).

Cloninger (1987) related his original Type I and Type II personalities to genetic studies of children adopted from or into families with an alcoholic parent. How did the genetic inheritance or the family environment predict risks of developing drink problems later on? The Type I pattern of risk, with a later onset of problems, was influenced strongly by the drinking patterns of *both* the biological and adoptive parents of boys and girls. Type II patterns with an early onset of problems related only to the *biological* father of boys (i.e. genetic factors seem to predominate in this group of alcoholic sons). Types I and II feel differently about their alcohol use, with Type I usually feeling a 'loss of control' during an extended period of drinking, followed by feelings of guilt and fear that they *cannot stop*, while Type II spontaneously seek alcohol with an 'inability to abstain' (i.e. they feel compelled to *start* drinking). This clinical distinction can be useful in understanding a client's motivation to continue drinking and these familial features have been confirmed in recent adoption studies (Sigvardsson *et al.* 1996):

- In adopted men with both genetic and environmental risk factors characteristic of Type I alcoholism, their lifetime risk of alcoholism is increased *fourfold* compared with others (neither the genetic nor the environmental factors acting on their own increase the risk of sons developing this type of alcoholism).
- In adopted men with the genetic risk factor characteristic of Type II alcoholism, regardless of their post-natal environment, their lifetime risk is increased *sixfold* compared with others.
- In both types of family, the majority of sons (over 80 per cent) *never develop alcoholism*. Factors unconnected with the parents (such as education, friends, limited occupational access to alcohol) presumably protect them from the causes of parental drinking.

To put the magnitude of genetic risks in perspective, consider the risks associated with occupation; for example, publicans running a bar have a *more than tenfold* increased risk of fatal alcoholic liver damage, compared with the general population. One such non-genetic factor which predicts lower levels of current alcohol intake and is also *protective* against developing alcoholism, which was discovered during epidemiological work with twins, is religious devotion (Kendler *et al.* 1997).

In the two adoption studies described above, identifying alcoholism depended on a medical diagnosis. For many years, it has been recognized that mood disorders like depression often co-existed with alcohol problems in some individuals, and Caan and Crowe (1994) observed that such patients had poor outcomes from psychiatric services. However, work for the 'Defeat Depression' campaign in Britain has uncovered many people, who had never sought help for either depression or drink problems from medical practitioners, who can be characterized as motivated to drink when their mood is low. In fact, having depression is the best predictor of those patients who will fail to attend any appointments made for them at an alcohol service (Shearer 1991) and, so, this sort of person who drinks in response to low mood is likely to be underrepresented in clinic populations! A spin-off of Schuckit's work on personality and alcohol use (Schuckit *et al.* 1997a) has been the discovery that about 15 per cent of the people who develop alcoholism in the USA had a depressive illness *before* their drinking problems began. Typically, these are married women who have never sought medical treatment.

Analysing the life stories of people who drink can help to understand what people find memorable about alcohol and how they perceive the motivation to continue drinking. Foxcroft and colleagues (1994) recorded this portrait in the words of one 15-year-old:

> I got drunk because I was having problems with my mum ... I like drinking because it helps me forget my problems.

At high doses of alcohol, memory systems break down altogether, but at lower doses we can easily observe 'state-dependent learning', where memories laid down when intoxicated are recalled only in a similar state of intoxication and, conversely, we have relative difficulty recalling events from the sober state when drunk. In narratives from three psychiatric outpatient clinics (W. Assin and W. Caan, unpublished), certain attenders told almost identical stories: periodically, they tended to ruminate all day on painful memories from early life which were associated with low feelings of helplessness and worthlessness, but over the years they had discovered that starting a drinking binge effectively blocked recall of these painful events, until over years a cycle of the same intrusive memories and drinking to forget them was established. Unfortunately, over time their mood between binges sank lower and lower, for which their worried patients failed to recognize any alcohol-induced contribution. What sort of painfully intrusive memories can relate to maladaptive drinking? Moncrieff and colleagues (1996) found 54 per cent of women in their alcohol services had been sexually abused as children (this is double the rate recently described in general psychiatry and about ten times the rate seen in patients with problems not related to mental health). Stories of abuse, desertion, homelessness or bereavement are common themes among young people who are beginning to drink heavily.

One of the hottest issues for mental health professionals is the phenomenon of *resilience*. Not every child with life experiences like desertion or bereavement subsequently develops a harmful alcohol habit. It is just now becoming apparent that psychosocial interventions can be effective in promoting mental health in high risk groups (Effective Health Care 1997; Health Education Authority 1997). The most common type of intervention for high-risk groups which seems effective in promoting coping is cognitive skills training. If, like a broken gramophone record, our thinking is stuck in the same groove, then fresh and productive cognitive skills might well be a good foundation for resilience.

Most previous textbooks emphasize how *bad* learning is under the influence of alcohol, with memory blackouts and other problems given prominence. I want to emphasize how *much* people learn while they are drinking. People can recall an accurate subjective picture of their early alcohol experiences which matches the objective measures made 15 years earlier (Schuckit *et al.* 1997b). In the course of drinking, we learn many environmental cues that precede and can eventually prompt drinking behaviour (Glautier 1996). The more often we drink, the more complex the chain of events which build up to alcohol consumption, but also the more stereotyped or automatic this sequence of cues and behaviours becomes the more we drink 'absent-mindedly' (Tiffany 1990) with little effort or decision making required. I once counted while a friend (who subsequently succumbed to liver disease) drank 32 units continuously in a stereotyped manner like a chain-smoker lighting cigarettes. The only variation in his sequence of movements was an unconscious move from one chair (in which he always drank cider) to another chair (in which he did his porter drinking). The benefit of understanding the cognitive structures that underpin bad habits, with their identifiable antecedents and beliefs, is that we can 'unlearn' these sequences that lead to drinking. Even for individuals with co-morbid depression, alcohol and illicit drug problems and impulsive criminal behaviours, the emancipation offered by Cognitive Behavioural Therapy is showing great promise (Caan 1995).

A test for your understanding thus far

The author's first really hazardous alcohol use, at the age of 8 years, involved stealing alcohol from an adult party combined on impulse with trying to climb a series of trees while intoxicated. Through adolescence, substance use was never continuous but spontaneous and reckless; for example, by 17 unpleasant experiences like a humiliating alcohol-related blackout during school did not deter subsequent drinking.

Within what common pattern of alcohol use do you think that behaviour fits?

Our genes and our environment interact, as we mature, either to promote or to protect against hazardous alcohol use in adulthood. For example, experiences of strictness at school amplify risk but family cohesion reduces risk, subsequently (Jang *et al.* 2001). The effects of drinking on an individual's mood can be dramatic: 94 per cent of incidents of self-harm in the Emergency Room study of Simpson *et al.* (2001) had recently been drinking. Across the world, there is a pattern of need for professionals concerned with suicide prevention to address the dangerous use of alcohol (Foster 2001).

Alcohol use seen 'from a distance': effects over time on individual lives and affected groups

Most adults develop drinking habits within the 'sensible' limits of consumption (although temporary ups and downs in alcohol use will accompany life events, such as an experience of unemployment or pregnancy). At low levels of drinking, they are more likely to experience some benefits from alcohol like the lubrication of social events than some physical, psychological or social harm. Quite often, however, some cumulative changes with repeated drinking can be seen. The most obvious change is that alcohol loses its anxiolytic and hypnotic effects as we become *tolerant* to its actions. In fact, sleep patterns tend to get worse with regular drinking and previously unknown experiences of irritability or anxiety appear, with night-time panic attacks an extreme example which the sufferers rarely attribute to alcohol. Kushner *et al.* (1999) followed up college students without a history of anxiety for 7 years, and those who had alcohol dependence at college were subsequently four times more likely to develop a clinical anxiety disorder, compared with the other graduates.

After a heavy bout of drinking, the production of copious acetaldehyde leads to some noxious metabolites appearing in the brain, including B-carbolines and isoquinolines (Caan 1996a). Compared with alcohol, B-carbolines have the opposite action on the GABA-A receptor, causing anxiety and insomnia, and in some cases an accumulating exposure to this toxin after frequent bingeing might cause panic attacks, hallucinations or seizures to appear (Caan 1996b). Compared with alcohol, isoquinolines have the opposite action on the NMDA receptor, and in rats this type of excitotoxin causes long-lasting damage to specific neurones, including the small number of strategically very important neurones which synthesize acetylcholine or 5-HT. Long *after* a period of heavy drinking has ended, these rats (compared with rats without a history of drinking) show poorer reasoning in solving unpredictable puzzles, impaired 3-D spatial memory and intrusive outbursts of aggression towards other rats who are trying to follow the same food trail (perhaps the animal equivalent of road rage). In human studies, Adrian Bonner (1996) has been trying to relate the action of chronic alcohol use in depressing 5-HT synthesis to compulsive alcohol seeking and antisocial behaviours using Cloninger's genetic typology. If some families have a tendency to both low rates of 5-HT synthesis and to impulsive behaviour, he predicts these persons will respond differently to 5-HT depletion by chronic alcohol use than people who start with normal levels of 5-HT activity in the brain. The gradual development of craving for alcohol is one of the long-running areas of controversy in alcohol studies (Tiffany 1990) and it is likely to involve more than one mechanism in different individuals. In mice exposed chronically to alcohol, the fits which occur after an abrupt withdrawal of alcohol can be prevented by drugs which block the L-calcium channels in the brain (Littleton *et al.* 1988). In mice and men, there is growing research interest in the role of these calcium channels in the development of cravings for alcohol. Since these voltage-sensitive calcium channels play an important role in regulating the release of both 5-HT and dopamine from neurones, there may be common brain systems relating the theories about alcohol seeking of Littleton, Bonner and Robbins, above.

As alcohol habits develop in adolescence, other behaviours are influenced in parallel. The series of *Teenage Health and Lifestyle* surveys by the Health Education Authority showed that the level of use of other substances (like tobacco) parallels the level of drinking, and there are also effects on diet and any active involvement in leisure pursuits. Gambling and drinking episodes can become entrained in the same individual, and Sue Fisher's (1996) work for the Office of the National Lottery showed that under-age gamblers were also likely to be under-age drinkers. Sometimes, long-term alcohol use brings completely new patterns of thinking. One striking example is the development of morbid jealousy (Michael *et al.* 1995). Typically, a husband begins manifesting suspicious, angry and unreasonable thoughts about his wife when he is intoxicated, but the overwhelming jealousy begins to extend into the periods when he is sober, and a pattern of violence against the supposedly unfaithful wife is likely to erupt within that relationship. In the forensic field, Chacksfield and Forshaw (1997) have found that issues around locus of control (the sense of whether we can choose our own actions or whether we feel compelled to act because of external pressures) are central to understanding the *interaction* of alcohol use and patterns of violence over time which are seen in the mentally disordered offenders at Broadmoor Hospital.

Using alcohol not only affects the person drinking but also the people who depend on them or care for them; for example, when Iwaniec (1995) reported on emotionally abused and neglected children, 75 per cent of their mothers 'seriously misused alcohol'.

If the non-drinking members of a family begin to organize all the important aspects of their lives around the drinker, they may develop 'co-dependence' on alcohol. Typically, a doting wife or mother spends years orbiting around the drinking which has become central to all the family relationships; this subordination of control 'to the drink problem' can bring its own continuing psychosocial disabilities for the surviving widow or child of an alcoholic even after the drinker is gone.

Mutually damaging behaviours could also occur between the paid home carers of elderly people and an old person with drink problems (Herring and Thom 1997). Given that a history of heavy alcohol use can predispose to both depression and dementia in later life, it is worth considering practice around alcohol use in the context of such caring relationships. A focus of recent training by the Mental Health Foundation (1998) has been to balance the 'care worker's duty of care' with respect for the rights of the older person 'to make their own lifestyle choices'.

Alcohol seen from the frontier of 'problem drinking'

When does drinking become 'a problem'? One approach is to ask community samples of people who drink to report any problems they themselves relate to alcohol. In the Canadian study of Bourgault and Demers (1997), 21 per cent of the drinking population reported one or more problems in the last 3 months, with harm most commonly reported in the financial and physical health areas. Dose and context of alcohol use predicted the risk of harm, with the most problems being reported by men who drank five or more drinks at one time, in a solitary setting. However, Grant's (1997) study of reasons why many Americans do not seek help for their drink problems found that the two most common reasons given were that they:

> 'Didn't think drinking problem was serious enough' or 'Thought it was something you should be strong enough to handle'.

Therefore, it is very important for professionals to begin 'asking the right questions' (Paton and Saunders 1988) to help their clients with concerns around alcohol to put potential problems in a realistic, mutually honest light. Paton and Saunders give a basic framework for discussing drink problems, and those readers who want a more detailed guide to engaging the person with hazardous drinking and recognizing the true severity of their problems may find useful the two *handbooks* of the Medical Council on Alcoholism (1990) for nurses or for general practitioners, which give lists of physical, psychological, familial and social indicators.

A single behaviour (such as drinking a lot) or a single harmful consequence of alcohol (like cirrhosis or the severe amnesia of Korsakov's psychosis) cannot define the nature of a person's progression to alcoholism. The WHO has adopted a 'syndrome' approach to defining the presence of an illness, 'dependence' on alcohol. The WHO describe seven dimensions of problem in dependence which tend to arise together and to develop together over time:

- a subjective awareness of compulsion to use a drug (*e.g. alcohol*) or drugs, usually during attempts to stop or moderate drug use;

- a desire to stop drug use in the face of continued use;
- a relatively stereotyped drug-taking habit (i.e. a narrowing in the repertoire of drug-taking behaviour);
- evidence of neuroadaptation (tolerance and withdrawal symptoms);
- use of the drug to relieve or avoid withdrawal symptoms;
- the salience of drug-seeking behaviour relative to other important priorities;
- a rapid reinstatement of the syndrome after a period of abstinence.

Individuals will show more severe or continuous problems in different dimensions of this syndrome, but some combination of biological, psychological and social damage is usually present in individual alcoholics.

For over 20 years, a useful tool in assessing whether a person with drink problems might be entering the realm of dependence has been the 4-item *CAGE* questionnaire:

Have you ever felt you should Cut down on your drinking?
Have people ever Annoyed you by criticizing your drinking?
Have you ever felt bad or Guilty about your drinking?
Have you ever had a drink first thing in the morning to steady your nerves or to get rid of a hangover (Eye opener)?

If clients answer 'yes' to two of these questions, they are possibly crossing the unmarked frontier into dependence and if they answer 'yes' to three or four items, it is very likely that a more detailed assessment will confirm alcohol dependence is present (Medical Council on Alcoholism 1990).

In Chapter 3, the properties of five common screening questionnaires, including CAGE, were described. A recent review (Saitz 1999) suggests that all these short questionnaires are widely acceptable, but that for specific populations some are more sensitive in detecting drink problems than others; for example, the most sensitive test for *current* hazardous drinking among North American women is the *TWEAK*, with its questions about Tolerance, Worried, Eye openers, Amnesia and 'Kut' down: giving two positive answers identifies 87 per cent of affected women.

The seven WHO dimensions have proved a very practical framework, but it is important to remember that the cultural concept of alcoholism can be broader than this dependence syndrome. If you are a pastor, alcoholism may present extra spiritual dimensions to someone's drink problems, if you are a probation officer a client's capacity to comply with the instructions of the court may be a key feature of their problems and if you are a teacher your student's ability to fulfil their learning role may colour your judgement of their 'alcoholism'. The *social handicap* consequent to an individual's drinking (i.e. the way *other significant people respond* to their drinking), may be a feature of progressive alcoholism which many caring professionals ignore at their peril; for example, I remember a senior manager of a multinational bank who was expected to entertain prospective new clients with expense-account drinks at all times of his working day as an essential part of his job. Shortly before his drawn-out death from liver disease, his bosses were still encouraging him to drink with customers; any true description of his 'alcoholism' had to include a social dimension as well as characteristics of the individual drinking.

Alcohol use seen in 'dependence'

First, alcohol dependence is a common phenomenon: the OPCS (1994) found about 8 per cent of men and 2 per cent of women in Britain were currently dependent, with most of this dependence unrecognized by the Health Service and by these addicts themselves. The peak age for finding dependence was not in middle age, it was aged 20–24 for men and aged 16–19 for women.

The psychological dimensions of dependence are not hard to spot. People feel compelled 'by the addiction' (i.e. with an external locus of control) to keep on drinking, when they try to cut down. They experience a distressing dissonance between wanting to stop and finding that the drinking remains an unending feature of their lives (so it is not surprising that feelings of guilt, anger or worthlessness deepen over time). The salience of alcohol preoccupies them; they have to plan trips or events to ensure a reliable supply of drink is available and, if unexpectedly offered an opportunity to have a drink, they tend to drop whatever else they had planned to do and have the drink first.

The good news is that dependence is not a static condition and a number of psychological treatments can contribute over time to an improvement in the condition of many people who are affected by alcoholism. Hodgson (1994) has reviewed the effectiveness of thirty-four recognized treatment approaches and six interventions (all of which have a basis in the effects of alcohol on the mind) that consistently give positive outcomes:

1   social skills training;
2   self-control training;
3   brief motivational counselling;
4   marital therapy (with a behavioural basis);
5   community reinforcement techniques (usually involving a spouse, employer or probation officer);
6   stress management training.

Alcoholism often generates feelings of hopelessness in clients and professionals, and there is a high mortality associated with this syndrome, but after years of ups and downs a high proportion of alcoholics achieve some measure of *independence* from alcohol. Factors like entry into some form of treatment (above), supportive relatives (preferably showing consistent 'tough love' rather than colluding with co-dependence) and employment in a supportive environment (preferably not as a publican) *promote* recovery. Learning from the ups and downs along the way and cultivating the ability to reflect on life's changes are definitely *assets* during rehabilitation from the damage done by such a chronic/relapsing condition as dependence on alcohol.

Alcohol seen from a 'sober' view 'in recovery' from dependence

Years or even decades of living with the multiple problems of the dependence syndrome (above) leave residual effects that are still noticeable long after sobriety has been gained. We often cannot undo the damage to relationships that happened during drinking, so many recovering alcoholics value the new relationships formed through mutual aid groups like AA and attend meetings long after taking their last drink. Problems with memory and anger management may persist for years after detoxification. Over about 2 months after starting a detox, brain metabolism shows definite improvements, but the metabolism of frontal, parietal and temporal cortex starts from a lower and lower baseline depending on how many years of drinking preceded treatment (Volkow *et al.* 1994). In some sober ex-alcoholics, CT scans and psychometric tests show continuing improvements up to about a year after detoxing (Caan 1996a). Even after years of abstinence, dependence can be rapidly reinstated (see the final bullet point of the syndrome on p. 63), and, so, an understanding of the immediate cues to drinking (Glautier 1996) and the automatic routines that anticipate drinking (Tiffany 1990) can help us to build strategies for relapse prevention. A valuable perspective on recovery is that it gradually shifts from preoccupations with a past life to creating the life ahead. The majority of people currently dependent on alcohol are under 30 years old (OPCS 1994). In undertaking rehabilitation from the 'scars' and 'ghosts' that can haunt sober times at the start of independence, youth is a great advantage in tending to look forward to life; in the words of Pope John Paul II (1991), at the first Vatican conference on alcoholism, 'hoping against hope'.

References

Alcohol Concern (1997) *Measures for Measures: A Framework for Alcohol Policy.* London: Alcohol Concern.

Alert (1994) 'Drunkenness involved in nearly half of all violent incidents', *Alcohol Alert*, **8**, 14–15.

Altman, J., Everitt, B. J., Glautier, S., Markou, A., Nutt, D., Oretti, R., Phillips, G. D. and Robbins, T. W. (1996) 'The biological, social and clinical bases of drug addiction: Commentary and debate', *Psychopharmacology*, **125**, 285–345.

Arjuna (1995 translation) *Krishna's Dialogue on the Soul*. London: Penguin.

Bonner, A. (1996) 'Molecules, brain and addictive behaviour', in A. Bonner and J. Waterhouse (eds) *Addictive Behaviour: Molecules to Mankind*, pp. 213–29. London: Macmillan.

Bourgault, C. and Demers, A. (1997) 'Solitary drinking: A risk factor for alcohol-related problems?', *Addiction*, **92**, 303–12.

Caan, W. (1992) 'Alcohol and HIV/AIDS: An overview of the research', in T. Bunyan (ed.) *Alcohol and HIV/AIDS*, pp. 11–17. Roehampton: Richmond, Twickenham and Roehampton HA (for South West Thames RHA).

Caan, W. (1995) 'Continuity of care key to dual diagnoses', *Nursing Standard*, 28 June–4 July, 42.

Caan, W. (1996a) 'Exogenous drugs and brain damage', in A. Bonner and J. Waterhouse (eds) *Addictive Behaviour: Molecules to Mankind*, pp. 56–68. London: Macmillan.

Caan, W. (1996b) 'Beta-carbolines in chronic alcoholics', *Addiction Biology*, **1**, 309.

Caan, W. and Crowe, M. (1994) 'Using readmission rates as indicators of outcome in comparing psychiatric services', *Journal of Mental Health*, **3**, 521–4.

Chacksfield, J. D. and Forshaw, D. M. (1997) 'Occupational therapy and forensic addictive behaviours', *Brit. J. Ther. Rehabil.*, **4**, 381–6.

Cloninger, C. R. (1987) 'Neurogenic adaptive mechanisms in alcoholism', *Science*, **236**, 410–16.

Deery, H. A. and Love, A. W. (1996) 'The effect of a moderate dose of alcohol on the traffic hazard perception profile of young drink-drivers', *Addiction*, **91**, 815–27.

De Moore, G. M. and Robertson, A. R. (1996) 'Suicide in the 18 years after deliberate self-harm – a prospective study', *Brit. J. Psychiat.*, **169**, 489–94.

Effective Health Care (1997) *Mental Health Promotion in High Risk Groups*, Vol. 3, Bulletin 3. York: NHS Centre for Reviews and Dissemination.

Engs, R. C. and Van Teijlingen, E. (1997) 'Correlates of alcohol, tobacco and marijuana use among Scottish postsecondary helping-profession students', *J. Studies on Alcohol*, **58**, 435–44.

Fisher, S. (1996) 'A preliminary study of underage spending on the national lottery', in *Social Research Programme*. London: OFLOT.

Foster, T. (2001) 'Dying for a drink. Global suicide prevention should focus more on alcohol use disorders', *BMJ*, **323**, 817–18.

Foxcroft, D. R. and Lowe, G. (1993) 'Self-attributions for alcohol use in older teenagers', *Addict. Res.*, **1**, 1–9.

Foxcroft, D. R., Lowe, G. and May, C. (1994) Adolescent alcohol use and family influences: Attributive statements by teenage drinkers. *Drugs: Education, Prevention and Policy*, **1**, 63–9.

Glautier, S. (1996) 'Controlling cues that trigger drug use', *MRC News*, **71**, 20–1.

Goodwin, D. W. (1994) *Alcoholism: The Facts*. Oxford: Oxford University Press.

Grant, B. F. (1997) 'Barriers to alcoholism treatment: Reasons for not seeking treatment in a general population sample', *J. Studies on Alcohol*, **58**, 365–71.

Health Education Authority (1997) *Mental Health Promotion. A Quality Framework.* London: HEA.

Herring, R. and Thom, B. (1997) 'Alcohol misuse in older people: The role of home carers', *Health and Social Care in the Community*, **5**, 237–45.

Hodgson, R. (1994) 'The treatment of alcohol problems', *Addiction*, **89**, 1529–34.

Holder, H. D. (1997) 'Alcohol use and a safe environment', *Addiction*, **92** (Suppl. 1), S117–20.

Iwaniec, D. (1995) *The Emotionally Abused and Neglected Child.* Chichester: Wiley.

Jang, K. L., Vernon, P. A., Livesley, W. J., Stein, M. B. and Wolf, H. (2001) 'Intra- and extra-familial influences on alcohol and drug misuse: A twin study of gene-environment correlation, *Addiction*, **96**, 1307–18.

Kendler, K. S., Gardner, C. O. and Prescott, C. A. (1997) 'Religion, psychopathology, and substance use and abuse: A multimeasure, genetic-epidemiologic study', *Amer. J. Psychiat.*, **154**, 322–9.

Kushner, M. G., Sher, K. J. and Erickson, D. J. (1999) 'Prospective analysis of the relation between DSM-III anxiety disorders and alcohol use disorders', *Amer. J. Psychiat.*, **156**, 723–32.

Littleton, J., Harper, J., Hudspith, M., Pagonis, C., Dolin, S. and Little, H. (1988) 'Adaptation in neuronal Ca^{2+} channels may cause alcohol physical dependence', in M. Lader (ed.) *The Pharmacology of Addiction*, pp. 60–72. Oxford: OUP.

Medical Council on Alcoholism (1990) *Hazardous Drinking: A Handbook for General Practitioners and Alcohol and Health: A Handbook for Nurses, Midwives and Health Visitors.* London: MCA.

Mental Health Foundation (1998) 'Alcohol and older people project', *MHF Briefing*, 21.

Michael, A., Mirza, S., Mirza, K. A. H., Babu, V. S. and Vithayathil, E. (1995) 'Morbid jealousy in alcoholism', *Brit. J. Psychiat.*, **167**, 668–72.

Moncrieff, J., Drummond, D. C., Candy, B., Checinski, K. and Farmer, R. (1996) 'Sexual abuse in people with alcohol problems. A study of the prevalence of sexual abuse and its relationship to drinking behaviour', *Brit. J. Psychiat.*, **169**, 355–60.

Nash, J. M. (1997) Addicted. *Time*, 26 May, 52–8.

OPCS (1994) *OPCS Surveys of Psychiatric Morbidity in Great Britain*, Bulletin 1. London: OPCS Social Survey Division.

Paton, A. and Saunders, J. B. (1988) 'Asking the right questions', in *ABC of Alcohol.* London: BMJ.

Platt, S. (1997) 'Message in a bottle', *Community Care*, 15–21 May, 18–19.

Pope John Paul II (1991) *Contra spem in spem. Drugs and alcoholism against life.* Rome: Pontifical Council for Pastoral Assistance to Health Care Workers.

Saitz, R. (1999) 'Review: Some alcohol screening tests have acceptable test properties for use in general clinical populations of North American women', *Evidence-Based Mental Health*, **2**, 20.

Samson, H. H. and Harris, R. A. (1992) 'Neurobiology of alcohol abuse', *Trends in Pharmacological Sciences*, **13**, 206–11.

Schuckit, M. A. (1994) 'Low level of response to alcohol as a predictor of future alcoholism', *Amer. J. Psychiat.*, **151**, 184–9.

Schuckit, M. A., Tipp, J. E., Smith, T. L., Shapiro, E., Hesselbrock, V. M., Bucholz, K. K., Reich, T. and Nurnberger, J. I. (1995) 'An evaluation of type A and B alcoholics', *Addiction*, **90**, 1189–203.

Schuckit, M. A., Tipp, J. E., Bergman, M., Reich, W., Hesselbrook, V. M. and Smith, T. L. (1997a) Comparison of induced and independent major depressive disorders in 2,945 alcoholics', *American Journal of Psychiatry*, **154**, 948–57.

Schuckit, M. A., Smith, T. L. and Tipp, J. E. (1997b) 'The self-rating of the effects of alcohol (SRE) form as a retrospective measure of the risk for alcoholism', *Addiction*, **92**, 979–88.

Shearer, C. M. (1991) 'Alcohol and drug users who seek help: Their motivations and expectations of treatment', unpublished MSc. thesis, University of Surrey, Guildford.

Sigvardsson, S., Bohman, M. and Cloninger, C. R. (1996) 'Replication of the Stockholm Adoption Study of alcoholism. Confirmatory cross-fostering analysis', *Arch. of General Psychiat.*, **53**, 681–7.

Simpson, T., Murphy, N. and Peck, D. F. (2001) 'Saliva alcohol concentrations in accident and emergency attendances', *Emerg. Med. J.*, **18**, 250–4.

Suominen, K., Isometsa, E., Henriksson, M., Ostamo, A. and Lonnqvist, J. (1997) 'Hopelessness, impulsiveness and intent among suicide attempters with major depression, alcohol dependence, or both', *Acta Psychiat. Scandinavica*, **96**, 142–9.

Tiffany, S. T. (1990) 'A cognitive model of drug urges and drug-use behaviour: Role of automatic and nonautomatic processes', *Psychol. Rev.*, **97**, 147–68.

Volkow, N. D., Wang, G-J., Hitzemann, R., Fowler, J. S., Overall, J. E., Burr, G. and Wolf, A. P. (1994) 'Recovery of brain glucose metabolism in detoxified alcoholics', *Amer. J. Psychiat.*, **151**, 178–83.

Volkow, N. D., Wang, G. J., Fowler, J. S., Logan, J., Hitzemann, R., Ding, Y. S., Pappas, N., Shea, C. and Piscani, K. (1996) 'Decreases in dopamine receptors but not in dopamine transporters in alcoholics', *Alcoholism: Clin. Experim. Res.*, **20**, 1594–8.

Chapter 6

The sociology of alcohol abuse

Alexander Baldacchino

Key points

- *Understanding the sociology of alcohol use enables us to place individual alcohol problems within a coherent context of history, family and community characteristics.*
- *This understanding draws on the varying patterns of drinking observed at different times and in different environments, and suggests avenues of intervention to reduce the problems related to alcohol use in a particular, cultural context.*

Sociology is the study of human social life, groups and societies. It provides us with an opportunity to step back from our own personal interpretations of the world and to look at the social influences which shape our lives. The intricate relationship between intoxicating substances and human behaviour on different levels of economics, politics, health, anthropology and history remind us of the extremely wide scope sociology offers to us, to help us analyse and understand 'addictive behaviour'. As therapists, understanding these sociological perspectives of alcohol misuse will help us become more objective and effective.

This chapter is on three sections. Section 6A looks at three aspects of drinking within a sociocultural model that will enable the reader to gain a better understanding of the associated alcohol-related behaviour, and ultimately enable the reader to appreciate these behaviours from the perspective of the problem drinker. Section 6B provides a theoretical insight into the sociological models of addictive behaviour. The concluding section (Section 6C) offers a case study of one community.

6A SOCIOCULTURAL ASPECTS OF ALCOHOL ABUSE

A sociocultural model is useful to:

a   Study the historic developments of alcohol abuse.

b   Identify and differentiate between different societies, cultural groups or subgroups on their views and attitudes towards alcohol problems in order to engage them better in a treatment modality.
c   Study overt and covert meanings of drinking and how society might perpetuate excessive or hazardous alcohol intake in order to plan educational interventions.

It is fundamental that, when using sociocultural models, we need to specify:

1   The level of culture being examined. Is the theory holding for any society or for a specific culture or subgroup within a society?
2   The level, setting and pattern of drinking. Is the theory aimed at reasons for drinking alcohol, alcohol-related disabilities, alcoholism or the alcohol dependence syndrome?

History of alcohol at a glance

The historical dimension helps us understand the distinctive nature of our world today if we are able to compare it with the past. Current debates on alcohol use and controls can be given a fresh sense of perspective by comparing them with the preoccupations of the past.

The use of alcohol can be traced back to the Neolithic age. Prehistoric cave paintings in the Mediterranean basin dated 8000 BC show that honey was sought after and stored to be fermented into an intoxicating drink called *mead*.

Beer and berry wine were known and used from about 6400 BC. The Egyptians discovered the basic process of brewing. Barley beer is still made today by the same methods they pioneered. The Egyptians also had their own goddess of beer, Menquet.

Hathor, the Sacred Bull was the divinity of the grapevine and was duly honoured on a monthly 'Day of Intoxication'. Grape wine was consumed from about 3000 to 4000 BC. A Persian myth tells of its discovery by a woman who first drank the fermented juices from grapes that had been stored in a pottery jar. The Old Testament condemns drunkenness but not alcohol. The 'strong wine' of the Bible was probably undiluted wine. Alexander the Great was one of many generals who was a heavy alcohol drinker. During the same period the Greek philosopher and founder of medicine, Hippocrates, was the first to describe the medical complications of alcohol.

Elsewhere, different cultures produced fermented drinks out of their own native plants: The Siberian cultures used red algae. The North American Indians made liquor from maple syrup. The Aztecs made *pulque* from agave and aqua cactus and the Amazon Indians used jungle fruits. Oriental nations still make alcohol from rice.

Throughout antiquity, it was not an unusual practice to mix additional intoxicants, such as mescaline, harmaline or other alkaloids, into the fermented drinks.

The term 'alcohol', which in Arabic means 'finely divided spirit', was probably first coined in about AD 800 when the process of distillation was discovered. The Italians introduced fine distilled wine for medicinal purposes in the thirteenth century and alcohol generally became the 'intoxicant of choice' in Western (Judeo-Christian) culture.

Whisky comes from the Irish Gaelic equivalent of 'aqua vitae' or water of life and was already commonplace around 1500. By the early 1800s, annual per capita consumption of whisky in the USA was 6–7 gallons per person. In the UK, drinking and drunkenness reached a peak during the gin epidemic between 1720–1750. It resulted in 15 per cent of all London adult deaths. The subsequent introduction of coffee and tea helped sober up the inebriated population. On the European front, *Absinthe* was a popular blue–green liquor in Paris in 1860. This was prepared from a distillate of wormwood and aromatic herbs and spices. The alcohol content varied between 45 and 75 per cent. It was strong enough to cause hallucinations and convulsions. Absinthe drinkers provided subjects for artists like Pablo Picasso, Edouard Monet and Edgar Degas. It also inspired the work of heavy drinkers like Vincent Van Gogh, Paul Verlaine and Amedeo Modigliani.

Sociocultural influences within specific subgroups

This anthropological dimension allows us to appreciate the kaleidoscope of different forms of human social life present. This appreciation teaches us more about the distinctiveness of our own specific patterns of behaviour. It facilitates a more effective therapeutic engagement with the problem drinker.

Broad cultural groups

Several cultural attitudes have been considered to be socially protective against alcohol misuse. The exposure of children to alcohol early in life, in the *context* of a strong family or religious group, with the alcoholic beverages considered mainly as a food and usually consumed with meals, has consistently shown to be protective. Other protective cultural attitudes include the presence of parents who consistently practise moderate drinking and who present the act of drinking alcohol as being morally neutral. This tends to encourage people to view alcohol as a drink with clear ground rules to define what are acceptable and unacceptable social rituals (Bales 1946, O'Connor 1975).

Race and nationality

Ethnic groups differ in patterns of drinking behaviour, in social functions and cultural expectations attributed to drinking, and in rates of drinking problems. Italians and Jews traditionally exhibit low rates of problem drinking due to the drinking norms that discourage heavy drinking and intoxication, and the use of alcohol as part of a meal or religious rituals (Simboli 1985). In contrast, the Irish and Russians tend to have a higher prevalence of drinking problems since drinking is seen as the necessary medium used for all social occasions.

In situations where ethnic groups undergo rapid social and cultural change (e.g. following immigration or a move to urbanization), drinking patterns may shift as a result of exposure to new sociocultural patterns. Simboli (1985) found that American–Italian immigrants have increased heavy drinking compared with their Italian cousins.

In the USA, Castro and his colleagues (1999) have stressed the need for addiction services to develop 'cultural competence', where the specific population profiles of substance use are addressed within the traditions and values of *each* distinctive ethnic group.

Religion

Islamic cultures, Mormon communities and Sikh religious organizations deny the acceptability of alcohol use under any circumstances. Complete abstinence is considered as a religious and moral duty of every believer. In contrast, drinking wine is an integral part of many Christian ceremonies and rituals. A change of context, such as immigration, can affect strong religious attitudes about abstinence (McKeigue and Karmi 1993). In this study of south Asian immigrants to the UK, there were higher alcohol-related morbidity rates for some Asian groups than for the general population. There was also a high consumption of alcohol (e.g. among Sikh males).

Subgroups within a similar sociocultural dimension

Even within one community, the role of alcohol and drinking practices can change amongst different subgroups; for example, this can determine the way alcohol use emerges among young people in the community, and the way differences emerge in adulthood between the drinking habits of women and men.

Alcohol problems among the young

An adaptation of Zucker's model (1979) in Table 6.1 helps us understand the influences on drinking behaviour. The four classes of influence can be divided into:

Table 6.1 Alcohol and the young: influences on drinking behaviour and eventual drinking behaviour

Sociocultural/Community influences
  Socio-economic status of family
  Ethnicity
  Religious influences
  Neighbourhood values and behaviour alternatives

Family of origin
  Parent personality and child-rearing patterns
  Drinking patterns and attitudes in family of origin

Peer influences
  Peer personality and
  Socialization influences
  Drinking patterns and attitudes

Intra-individual influences
  Cognitive structure including attitudes, beliefs and values
  Personality structure like traits, motivation and temperament
  Genetic influences

1   sociocultural and community influences;
2   family of origin;
3   peer pressure;
4   intra-individual and personality aspects.

The model proposes that these four classes of influences have differential effects during the different developmental periods (e.g. family of origin during early childhood and peer pressure being more influential during adolescence).

We must not fall into the trap of stereotyping youth drinking as necessarily deviant and liable to increase violent behaviour. Information about drinking has to be analysed with a knowledge of regional and temporal variations in drinking (and other behaviour patterns) among the young people involved. In North-West England, a survey of 14 to 15-year-olds found many had already established a regular habit of using alcohol, with 30 per cent drinking every week and consuming 8 units of alcohol during a drinking episode (Newcombe *et al.* 1995).

Gender differences and alcohol misuse

There are significant differences between men and women who are hospitalized due to alcohol-related medical problems (Rimmel 1971).

Compared with women, men show an earlier onset of alcohol problems since males tend to start drinking alcohol at a younger age. Men progress more rapidly to daily drinking, show an increased incidence of dependency and a history of truancy and school problems. Subsequently there is an increased risk of loss of job and friends because of drinking and a history of alcohol-related arrests.

Female alcoholics tend to drink more often at home, alone or with a spouse who is also a heavy or problem drinker, to have shorter drinking bouts and to use alcohol more often to improve their performance at home or at work. They tend more often to perceive their alcoholism as becoming 'worse' and to combine alcohol more often with other substances such as tranquillizers. Female patients are more often diagnosed as suffering from 'depression' *rather* than an alcohol-related disorder, even though female drinkers characteristically show a higher prevalence of depressive illness than male drinkers from the same community (Golding *et al.* 1993, Horn *et al.* 1973).

As well as the gender differences in hospital samples (see p. 73), above, men and women in the community show differences in the relationship between changes in social roles and in drinking; for example, a longitudinal study of Dutch women who were single and drinking heavily at the beginning found they significantly reduced the frequency of their drinking bouts if they gained a marital role during the study (Hajema and Knibbe 1998). Gaining or losing marital roles did not affect the drinking behaviour of men in the same Dutch community.

The observations on the specific groups noted above are not exclusive and cannot be generalized to other subgroups of excessive drinkers who might be studied in other settings. The data might produce different conclusions.

The environmental factors relevant in the aetiology of alcohol misuse (Table 6.2)

The availability of alcohol is undoubtedly a fundamental and necessary environmental factor for the development of alcohol misuse and alcohol dependency. Therefore, factors that increase or reduce availability of alcohol have a direct relationship to the incidence of alcoholism (see Chapter 1). This relationship is true across a wide range of countries at almost every time in history.

6B THE SOCIOLOGICAL APPROACHES TOWARDS ALCOHOL ABUSE

Three different theoretical frameworks will be outlined below within which alcohol use can be conceptualized.

Functionalism approach

Social problems result from the instability caused by social change. At an individual level, the origins of the problematic behaviour are attributed to

---

Table 6.2 Environmental factors relevant in the aetiology of alcohol misuse

Socio-economic and political factors
Cost
Income
Legislation
Advertising practices
Drink-driving policies
Pub closing hours

Cultural factors
Induced tensions and needs for adjustment
Attitudes towards drinking
Availability of alternatives to drinking as means of adjustment

Occupational factors
Ready availability of alcohol at work
Frequent absence from home with social pressures to drink at work
Lack of supervision and social isolation
Long and irregular working hours
Very high or very low income
Exceptional stresses in the workplace

Family and immediate environmental factors
Psychological effects (e.g. modelling from parent and older siblings)
Emotional deprivation and abuse
Absent or chaotic family rituals

Presence of positive and negative life events
Bereavement
Change in job
Moving home
Winning the lottery
Separation and divorce

---

inappropriate socialization. There is, therefore, a conflict between the internalized goals and the opportunities available within the system to help achieve these goals. This is known as *anomie* or normlessness. This conflict can precipitate a breakdown in society. Merton (1971) identifies a set of four deviant responses to anomie:

1 *Innovation:* condoning of illegitimate means for achieving our goals.
2 *Rebellion:* challenging the sociocultural norms and values of a society by adopting a new set of values and priorities.
3 *Ritualism:* persisting in the pursuit of legitimate symbols of success such as hard work or educational achievements. However, the impetus behind it is not based upon the values that society incorporates but is directed by a hollow and dejected attitude empty of meaning or motive. The person merely goes through the motions.

4   *Retreatism:* abandoning not only the dominant cultural goals but also the institutionalized, approved means of achieving these goals. As a result, the drinker drifts into a homeless and vagrant state with no interest in rejoining the mainstream of society.

Functionalism is a sociological perspective that uncritically supports the status quo. The alcoholic is, therefore, perceived as a second-class citizen who failed and is struggling to survive within an existing social order that is in itself rigid and unable to integrate marginalized people. It ignores how society defines its 'problems' and the issue of homelessness as a living example of this approach; for example, the cross-cultural study in Romania and England by Margarit and Brandon (1999) did not find that homeless alcoholics 'abandoned' their homes or 'retreated' from their families but, rather, that others actively evicted them or families were broken up, because of their drinking behaviour.

Conflict approach

Society does not offer people equal access to wealth and to economic and political power. It is this *alienation* which is seen as the catalyst for the generation of social problems. Within this framework, any type of substance abuse is seen as either a form of escape or a resistance by members of subordinate social groups to this alienation. Such behaviour constitutes a threat to the control over labour power. The social establishment responds by defining excessive alcohol consumption or other forms of substance abuse as an illness or criminal offence needing medical treatment or legislative restrictions so as to remove, treat or punish unproductive human assets. Marxist ideology concluded that this alienation is a result of inequalities inherent in a capitalist system.

Whatever the origins of alienation, alcoholics' perception of their life as a whole is highly dependent on their social environment, with loneliness and pessimism markers of a very poor quality of life (Foster *et al.* 1999).

This approach fails to appreciate the importance of social processes that generate alcohol-related problems, or to consider the cultural and symbolic aspects of alcohol. An interesting development (Rotter 1992) has been the attempt to link the sociological concept of alienation with the psychological concept of locus of control derived from social learning theory (see Chapter 17 on rehabilitation).

Interactionist and constructionist approaches

These sociological perspectives are more concerned with processes rather than structures, meaning rather than causality:

1   *Interactionists* tend to focus on micro-settings such as neighbourhoods, schools and psychiatric wards and the interaction between individuals in these settings; for example, excessive alcohol and marijuana consumption with the associated behaviour is examined by Young (1973) and Becker (1967) from the standpoint of the participant, where society's reaction amplifies the problem.

2   *Constructionists* take a macro-view and target their investigation on the behavioural processes and related perceptions which lead individuals to create certain patterns of human interaction with a problematic *label*, leading to claims of deviance. An example is the justification to suppress certain political and cultural groups in the USA as a result of an alleged epidemic of heroin use in the 1980s.

Overall both approaches view substance abusers with sympathy, victimized and labelled by agencies of social control. Unfortunately, they do not lead to commitments to tackle the causes of deviant behaviour and can easily finish up building a conspiracy theory.

The three sociological theories described above try to help us understand deviance and conformity; culture and social interaction; structures of power and social institutions – with personal relevance to alcohol. Ideally, we need to combine the analytic strengths of the conflict and interactionist approaches with the focus of functionalism.

6C A CASE STUDY: THE MALTESE SCENARIO

The situation of the Maltese population concerning alcohol is better understood by examining various sociocultural interactions. This will help us understand what the Maltese perceive as a serious alcohol problem and the limitations of any interventions offered.

Figures on annual alcohol importation showed a rough estimate of 2.5 million litres of alcohol beverages. Although there are fourteen wine producing factories and co-operatives in the Maltese Islands, with a value output and sales of about £15 million during 1988, the practice of producing wine at home is also widespread. These figures can only shed light on a behaviour pattern that forms an integral part of the Maltese way of life. Malta is a country with a Latin culture where alcohol is socially and traditionally accepted. Drunkenness is generally only tolerated during the village or town *festa*. Beyond these festive days, people abusing alcohol become targets of malicious gossip and even overt hostility.

The local law enforcement agencies do not officially consider excess drinking when driving as a criminal offence. Advertisements of alcoholic beverages are common in both media and entertainment venues. The rate of admissions with a diagnosis of alcohol dependence syndrome has been progressively increasing with an estimated lifetime incidence for cirrhosis at 8.15 per 100,000 population.

> 1 Could you suggest what other factors might be relevant in the sociocultural study of alcohol abuse in the Maltese Islands?
> 2 Could you suggest another small culture/subgroup that you can relate more closely to?
> 3 To what extent do you think sociocultural influences are important in your own work setting or when you are seeing individuals with alcohol-related problems?

Hints: (1) Malta is a small country with a unique history and a tradition of strong family ties. (2) This culture could be a community you are living in now or one you were brought up in. It could also be a sector within your community with a predominant ethnic minority.

References

Bales, R. (1946) 'Cultural differences in rates of alcoholism', *Quart. J. Studies on Alcohol*, **6**, 489–99.

Becker, H. S. (1967) *Social Problems: A Modern Approach*. New York: Wiley.

Castro, F. G., Proescholdbell, R. J., Abieta, L. and Rodriguez, D. (1999) 'Ethnic and cultural minority groups', in B. S. McCrady and E. E. Epstein (eds) *Addictions: A Comprehensive Guidebook*, pp. 499–526. New York: Oxford University Press.

Foster, J. H., Powell, J. E., Marshall, E. J. and Peters, T. J. (1999) 'Quality of life in alcohol-dependent subjects – a review, *Quality of Life Research*, **8**, 255–61.

Golding, J. M., Burnam, N. A., Benjamin, B. and Wells, K. B. (1993) 'Risk factors for secondary depression among Mexican-Americans and non-Hispanic Whites: Alcohol use, alcohol dependence and reasons for drinking. *J. Nerv. Mental Dis.*, **181**, 166–75.

Hajema, K-J. and Knibbe, R. A. (1998) 'Changes in social roles as predictors of changes in drinking behaviour', *Addiction*, **93**, 1717–27.

Horn, J. L. and Wanberg, K. W. (1973) 'Females are different: On the diagnosis of alcoholism in women', paper given at Proceedings of the *First Annual Alcoholism Conference of the National Institute on Alcohol Abuse and Alcoholism*, DHEW publication No. (NIH) 74-675, Washington, DC: US Government Printing Office.

McKeigue, M. P. and Karmi, G. (1993) 'Alcohol consumption and alcohol-related problems in Afro-Caribbeans and South Asians in the United Kingdom', *Alcohol and Alcoholism*, **28**, 1–10.

Margarit, V. and Brandon, D. (1999) *Homeless in Romania*. Cambridge: Anglia Polytechnic University.

Merton, R. K. (1971) 'Social problems and the sociological theory', in R. K. Merton and R. Nisbet (eds) *Contemporary Social Problems*, 3rd Edn, New York: Harcourt Brace Jovanovich.

Newcombe, R., Measham, F. and Parker, H. (1995) 'A survey of drinking and deviant behavior among 14/15 year olds in North West England', *Addiction Research*, **2**, 319–41.

O'Connor, J. (1975) 'Social and cultural factors influencing drinking behaviour', *Irish J. Med. Sci.* (Supplement), 65–71.

Rimmel, J. (1971) 'Alcoholism II: Sex, socio-economic status and race in two hospitalized samples', *Quart. J. Studies of Alcohol*, **32**, 942.

Rotter, J. B. (1992) 'Cognates of personal control: Locus of control, self-efficacy, and explanatory style. Comment', *Appl. Prevent. Psychol.*, **1**, 127–9.

Simboli, B. J. (1985) 'Acculturated Italian-American drinking behavior', in L. Bennet and G. M. Ames (eds), *The American Experience with Alcohol: Contrasting Cultural Perspectives*, pp. 61–7, New York: Plenum Press.

Young, J. (1973) 'The amplification of drug use', in S. Cohen and J. Young (eds) *The Manufacture of News: Social Problems, Deviance and the Mass Media*. London: Constable.

Zucker, R. A. (1979) 'Developmental aspects of drinking through the young adult years', in H. T. Blane and M. E. Chafetz (eds), *Youth, Alcohol, and Societal Policy*. New York: Plenum Press.

Chapter 7

Alcohol and society

E. Jane Marshall

Key points

- *In order to understand the social impact of alcohol, it is important to distinguish between the popular 'myths' about drinking and the scientific observations which are available about alcohol consumption and its hazards.*
- *In particular, it is possible to investigate the impact of alcohol on the family and on the workplace.*
- *Different individuals and groups may differ in the degree of vulnerability or resilience they show in the face of hazardous alcohol use.*
- *Where harm does occur, it may follow a common pattern (e.g. damaged parent–child relationships in the family or impaired reliability and safety at work).*

Introduction

Alcohol is 'our favourite drug' (Royal College of Psychiatrists 1986). A legal and universally available substance, it gives great pleasure when consumed in moderate quantities, but also has significant toxic and pharmacological effects. Excessive consumption contributes to a wide range of physical, psychological and social problems at the individual level. It is estimated that 33,000 deaths per year in Great Britain are alcohol related (Alcohol Concern 1997), and that 15–30 per cent of male and 8–15 per cent of female admissions to general hospitals have alcohol-related problems (Chick 1994). Alcohol is a risk factor for many disorders, including hypertension, cardiac arrhythmias, stroke, coronary heart disease, cancers of the oesophagus, pharynx and larynx and the female breast, and cirrhosis (Anderson 1995). In many conditions, there is evidence of a dose–response relationship between consumption and risk. There is also evidence of a protective effect against coronary heart disease, but much of this effect can be achieved at consumption levels of less than 1 unit (8–10 g) daily, and is only relevant to men over 40 years and post-menopausal women.

Alcohol also contributes to a wide range of problems in society. It is involved in all types of accidents and contributes to 15 per cent of traffic deaths. Road traffic accidents in which alcohol is involved are more serious than accidents in which it is not (Glucksman 1994). Alcohol is also implicated in 26 to 54 per cent of home and leisure injuries (Edwards *et al.* 1994). It is associated with domestic violence, child abuse, crime, homicide and suicide. Up to 14 million working days are lost annually to alcohol-related problems. Alcohol costs the National Health Service £150 million per year (Alcohol Concern 1997). The cost in terms of human suffering is incalculable.

Alcohol-related problems are not confined to individuals with chronic excessive alcohol consumption or those with alcohol dependence. Heavy social drinkers, particularly binge drinkers, are also at risk. Indeed, epidemiological evidence supports the view that most alcohol-related harm in the general population occurs in heavy drinkers who might not be considered to have an alcohol problem.

The *Oxford English Dictionary* defines society as: '*the state or condition of living in association, company, or intercourse with others of the same species; the system or mode of life adopted by a body of individuals for the purpose of harmonious existence or for mutual benefit, defence.*'

This chapter focuses on the ways in which alcohol affects the 'state or condition of living' as described above. Alcohol consumption and variations in alcohol use in different subgroups are considered. Certain myths about alcohol are addressed. The influence of gender, ethnicity and social groups (the family, peer group, friendship and occupational networks) are also considered, along with the issue of alcohol throughout the different stages of life.

Alcohol consumption

Historical data confirm that there have been considerable fluctuations in alcohol consumption throughout the ages, together with shifts in beverage choice (Edwards *et al.* 1994). In Britain, alcohol consumption increased in the early part of the nineteenth century and this was largely explained by an increase in the consumption of wine and spirits. Licensing controls were introduced in 1872, and the level of alcohol consumption declined after the mid-1870s (Plant 1997). Per capita alcohol consumption continued to fall until the 1930s. After the Second World War, consumption increased steadily, almost doubling between 1950 and 1976. There has been some levelling off in consumption since the early 1980s, reflecting the dominant pattern of consumption in Western Europe and North America (see Table 7.1). At the same time, consumption has increased markedly in Eastern Europe and Russia.

Table 7.1 Per capita alcohol consumption (litres of ethanol) in OECD countries 1970–1990

	1970	1980	1990
Australia	8.1	9.6	8.4
Austria	10.5	11.0	10.4
Belgium	8.9	10.8	9.9
Canada	6.1	8.6	7.5
Denmark	6.8	9.1	9.9
Finland	4.4	6.4	7.7
France	16.2	14.9	12.7
Germany	10.3	11.4	10.6
Great Britain	5.3	7.3	7.6
Iceland	3.2	3.9	3.9
Italy	13.7	13.0	8.7
Japan	4.6	5.4	6.5
Luxembourg	10.0	10.9	12.2
The Netherlands	5.6	8.8	8.2
New Zealand	7.6	9.6	7.8
Norway	3.6	4.6	4.1
Portugal	9.9	11.0	9.8
Spain	11.6	13.6	10.8
Sweden	5.8	5.7	5.5
Switzerland	10.7	10.8	10.8
Turkey	0.5	0.7	0.6
United States of America	6.7	8.2	7.5

Alcohol consumption is determined by a complex interaction between a variety of factors: economic, dietary, social and cultural. Long-term fluctuations in alcohol consumption are relatively common across countries, despite economic and cultural differences. In 1993, the United Kingdom per capita consumption was 7.0 litres of absolute alcohol (compared with 5.3 in 1970 and 7.3 in 1980). Similar levels of consumption were reported in Ireland, New Zealand, Canada and the USA (Brewers Society 1995). In France, Portugal and Spain, corresponding 1993 levels of consumption were somewhat higher at 12.3, 11.4 and 9.7 litres, respectively.

Individuals vary enormously in the amounts they consume. Some drink nothing at all, some drink only on special occasions and others drink very heavily; for example, over 50 per cent of women and nearly a third of men in Northern Ireland are abstainers, compared with 12 per cent of women and 7 per cent of men in Britain (Foster *et al.* 1990). Regional variations in levels of alcohol consumption have been identified in England. The highest prevalence of drinking more than 21 units per week among men was in the Northern Health Region (38.5 per cent) where mean consumption was 22.7 units per week. The lowest prevalence was in the North East Thames Health Region (23.5 per cent) where the mean consumption was 14.3 units per week (Colhoun *et al.* 1997). This report also showed that higher than average consumption among 'moderate' drinkers was associated with higher rates of heavy drinking and problem drinking.

An individual's risk of becoming a heavy drinker depends on the 'wetness' of the culture to which he belongs. (The term 'wetness' is used to denote a heavy drinking culture and the term 'dryness' to describe a culture in which alcohol is used sparingly.) Drinking habits are also influenced by the drinking habits within personal, social networks (Skog 1980). Thus, the drinking habits of a rugby player will be very different from those of a Methodist minister. Individuals who drink heavily are especially likely to smoke tobacco, to use illicit drugs and to engage in other forms of potentially high-risk behaviour. There is a genetic component to drinking and drinking behaviour. It is the manner in which this genetic vulnerability interacts with environmental factors which determines whether or not an individual develops a drinking problem.

Variations in alcohol consumption

Ethnicity

In the UK, Irish drinkers are subject to the stereotype of the 'ubiquitous drunken male labourer' (Greenslade *et al.* 1995). This stereotype is a distortion of reality, because per capita alcohol consumption in Ireland is low compared with other European countries (Conniffe *et al.* 1996), possibly related to higher rates of abstinence. Much Irish drinking is convivial, having social rather than religious connotations (Thom 1993). The cultural background and experience of Irish immigrants in the UK and the USA and their tendency to drink together in 'Irish' pubs interact to create a powerful social environment for drinking. In comparison, the use of alcohol in Jewish culture is bound up with religious symbolism from an early age; intoxication is abhorred. These factors foster the development of stable patterns of drinking and Jews have low rates of alcohol-related problems.

Major changes have occurred in Russia since the break-up of the Soviet Union, not least a significant decrease in life expectancy between 1990 and 1994 (Leon *et al.* 1997). This fall in life expectancy did not affect all parts of the country equally, but was more prominent in some of the wealthiest regions, suggesting that impoverishment was not the main explanation. Alcohol has played a major part in this decline in life expectancy, with binge drinking, in particular, being responsible for an excess of cardiovascular deaths in young men (Walberg *et al.* 1998). It seems that the effects of rapid social and economic change were felt most by young and middle-aged men, and that this interacted powerfully with a fall in the price of alcohol and a long cultural tradition of heavy drinking to promote a destructive pattern of binge drinking (Walberg *et al.* 1997).

Women

Social and cultural factors exert a powerful influence on the pattern and degree of drinking in women. There is still less social pressure on women to begin drinking and more pressure for them to stop. Women have, nevertheless, increased their alcohol consumption over the past 30 years. They are increasingly economically independent and are targeted by the drinks industry and advertising. Attitudes towards women drinking in public places have changed and more young women are drinking. The proportion of women drinking more than 14 units per week in Britain increased from 13 to 15 per cent between 1994 and 1998. This was accounted for by increases amongst women in the 18–24 year age band (Office of National Statistics 2000).

Young people

Typically, boys drink more than girls and older adolescents more than their younger counterparts. Most young people have tasted alcohol by the age of 13 and are familiar with 'designer drinks'. The frequency of drinking amongst young people increased throughout the 1990s and alcohol use is associated with smoking, illicit drug use and early sexual activity. Problems usually occur during or following episodes of intoxication. Drink driving is a serious problem amongst 17- to 24-year-olds, and road traffic accidents involving this age group are the highest cause of death. Complications also include other accidents, problems with parents and at school, and physical complications. Alcohol may be used to self-medicate anxiety, depression and to cope with eating disorders, and is a major risk factor for suicide in young people (Fombonne 1998).

Myths about alcohol

Any discussion on alcohol and society merits consideration of the following myths (Edwards *et al.* 1994):

Myth No. 1: 'Heavy' drinkers are a different species from 'ordinary' drinkers

Alcohol consumption is viewed on a continuum from no or light consumption through to heavy consumption. There is no clear boundary between normal and heavy drinking, and, although heavy drinkers are more at risk of developing alcohol-related problems, they are not a different species.

There is a genetic component to drinking behaviour and drinking problems (Edwards *et al.* 1997). The interaction of genetic vulnerability with environmental factors determines whether a certain person will go on to develop a drinking problem.

Myth No. 2: The prevalence and incidence of alcohol problems have nothing to do with population levels of drinking

Extensive research has shown that the higher the average consumption of alcohol in the population, the higher that population's incidence of alcohol-related problems will be. This holds true for almost all types of alcohol-related problems, including drink-driving offences, mortality due to cirrhosis and crimes of violence (Edwards *et al.* 1994). This relationship also holds true at the individual level. Thus, the risk of developing breast cancer in women, or alcoholic liver disease, or the risk of developing alcohol dependence, all vary in proportion to the individual's level of alcohol consumption (Edwards *et al.* 1994, 1997).

Myth No. 3: Drinking affects only a minority of the population

Heavy drinkers are at greater risk of developing alcohol-related problems and 'normal' drinkers rarely develop such problems. However, there are many more 'normal' drinkers than there are heavy drinkers in any population, and the majority of alcohol problems occur in the 'normal' drinkers, not in the heavy drinkers (Edwards *et al.* 1997). This is termed the 'preventive paradox', because, paradoxically, preventive measures to reduce alcohol-related problems must be directed at the whole population of drinkers, not just the heavy drinkers. The preventive paradox applies to some types of alcohol-related problems and not to others. The extent to which it applies depends on the proportion of drinkers affected by a particular problem at differing levels of alcohol consumption (Edwards *et al.* 1994). Thus, it applies to social problems such as drinking and driving, but does not apply to alcoholic cirrhosis, where only chronic heavy drinkers are at risk.

Myth No. 4: Acute, adverse consequences and alcohol-related accidents are not an issue

Table 7.2 shows that a significant proportion of road traffic accident fatalities involve alcohol (Edwards *et al.* 1994). Alcohol is also a major contributing factor to deaths from falls, from fire and drowning, as set out in Table 7.3. Data from Canada, Chile and the USSR (Table 7.4) show that alcohol consumption is associated with deaths from suicide and homicide.

Table 7.2 Proportion of fatal road traffic accidents involving alcohol (from Edwards et al. 1994)

Country	Year	%
Canada	1990	36.4*
USA	1990	50
France	1984	40*
Germany	1992	19
Sweden	1991	7.1–11.5
Chile	1991	50
Papua New Guinea	1981	50

* Figures for Blood Alcohol Concentrations > 80 mg %

Table 7.3 Involvement of alcohol in accidents other than road traffic accidents (from Edwards et al. 1994)

Causes of death	Canada (%)	USA (%)
Accidental falls	40	28
Fire	30	47
Drowning	30	34

Table 7.4 Involvement of alcohol in suicide and homicide

	Suicide (%)	Homicide (%)
Canada	30	60
Chile	38.6	48.6
USSR		>60

Alcohol and the family

Children

Families play an important role in passing on attitudes and beliefs about alcohol. Children do not usually have any direct experience of alcohol outside the home, so their parents' drinking habits and behaviour will have a lasting influence on them. If, as in France, where adults drink a glass of wine with their evening meal and do not become intoxicated, children soon learn to associate alcohol with food and not with intoxication. In other cultures (e.g. the Irish), drinking is done outside the home, and young children may not understand why or how their father (or mother) has 'changed'. The risk is that they will blame themselves.

The effect on a child of being brought up in a family where one or both parents have a drinking problem is incalculable. Certain 'resilience' factors may attenuate the ill effects. These include the degree of emotional support provided by either parent when sober and a variety of other emotional and social supports such as friends, extended family, school, church, etc. The age of the child when the parent develops the drinking problem and the personality of the child are also critical. A substantial number of families with alcoholic members remain intact, are economically viable, avoid dramatic and devastating family violence and suffer levels of anxiety and depression similar to those in the normal population. These families 'make do' and lead compromised lives (Steinglass *et al.* 1987).

Incessant alcohol-related arguments between parents, domestic violence and child abuse all cause emotional suffering which cannot easily be quantified. This suffering has lasting effects on personality development and the ability to form trusting relationships.

The effects on the family may be more indirect, taking the form of financial hardship (due to loss of job or spending income on alcohol), divorce, death of a parent, being taken into care. Children from 'alcoholic' families are prone to anxiety and low self-esteem (Zeitlin 1994). Girls are more likely to have a higher incidence of depression in childhood and adolescence, and boys to have a raised incidence of antisocial behaviour. They perform poorly at school and are highly likely to get into trouble with the law. These children are themselves at greater risk of developing a drinking problem in adulthood.

Child abuse

All forms of child abuse, emotional, physical and sexual, are associated with heavy drinking in parents. The explanation usually given is the disinhibiting effect of alcohol, usually in the father. However, the abuser is often someone other than the parent. This, together with the finding that father–daughter incest is associated with maternal drinking, suggests that impaired maternal protection and parental unavailability are important factors. Survivors of childhood sexual abuse are at risk of developing alcohol problems and typically start drinking at an early age.

Organization of the 'alcoholic' family

Alcohol use constricts family life, often in subtle ways, altering daily routines, family rituals and family problem solving. 'Alcoholic' families become organized so as to blunt the destabilizing effect of the 'alcoholic' parent, and act to blunt any potentially destabilizing event in family life. Normal developmental issues, such as adolescence, change of school, career issues, are ignored at the family level, and family members learn to develop an emotional distance and to fend for themselves (Steinglass *et al.* 1987). Families with an alcoholic member report that alcohol has affected their lives adversely, but cannot point to specific adverse effects (Clark and Midanik, 1982).

Foetal alcohol syndrome (FAS) and post-natal development

Heavy alcohol consumption during pregnancy is associated with harm to the developing foetus. Typical features of FAS include central nervous system dysfunction, abnormal facial features, behavioural deficits and growth deficiency. The incidence of FAS has been estimated to be 0.33 cases per 1,000 live births (Abel and Sokol 1991). A spectrum of milder alcohol-related birth defects, termed Foetal Alcohol Effects (FAE), occurs at lower levels of alcohol intake. Women who drink during pregnancy also have increased rates of complications of pregnancy and delivery, of spontaneous abortion and stillbirth. It has proved difficult to determine a level of drinking in pregnancy below which foetal development is unaffected.

Alcohol and work

Alcohol misuse contributes to significant absenteeism from work and poor performance on certain days (e.g. Monday mornings following a heavy drinking weekend). Absenteeism rates are most likely underreported with hangovers and alcohol-related disorders being self-certified by employees as 'colds', 'flu', 'gastric problems', 'food poisoning'. The cost of alcohol-related poor work performance is hard to assess, but, in the UK in 1998, the Trade Union Congress and Alcohol Concern produced an estimate of £3 billion. Alcohol contributes to a significant proportion of workplace accidents.

Excessive drinking has far-reaching effects on work performance at every level in the employment hierarchy. Certain jobs demand sobriety for reasons of public safety. Bus drivers, train drivers, pilots, ships' captains are thus subject to rigorous supervision. Drinking problems in other professions such as medicine, teaching, the Church, law and business throw up issues specific to these professions.

Alcohol and crime

The relationship between alcohol and crime is a complex one to which simple cause and effect explanations cannot be applied. Alcohol use and criminality have separate sets of predisposing factors, some of which may overlap. Some problem drinkers never become involved in criminal behaviour whereas some occasional, binge drinkers do. Alcohol-related crime can be classified as 'under the influence' (either while intoxicated or during the withdrawal period), acquisitive crime (including prostitution) and domestic violence/abuse. Violent crime is associated with intoxication and its attendant aggression. An Australian study (Stevenson et al. 1999) found that local rates of violent assaults in different communities could be predicted by the overall alcohol sales in the same localities. The state of intoxication at the time of a crime does not prove that alcohol caused the crime, but, rather, that it played a direct part in it. The study of alcohol-related crime has focused largely on drink-driving, public disorder, violence (including sexual abuse) and acquisitive crimes (thefts, burglary and shoplifting); for example, a survey in 1999 by *Community Care* found that every week 3 per cent of social workers suffer a violent assault at work, and Thompson (1999) reported that 'drink or drugs' were involved in 76 per cent of attacks on social workers. A large number of children may be harmed by a relatively small number of persistent paedophilic sex offenders. Raymond *et al.* (1999) assessed 45 of these child abusers in detail and found that 51 per cent had a history of alcohol disorders.

Other, minor, criminal acts may also be associated with alcohol use. These include travelling on public transport without a ticket, failing to pay for a meal in a restaurant, begging, urinating in the street and parking offences (Edwards *et al.* 1997).

Conclusion

This chapter does not pretend to be an exhaustive account of alcohol and society. Clearly, alcohol permeates society and plays a major role in the way humans interact with one another. Alcohol is a source of enjoyment but it also ruins lives, and many of us know at least one person whose life has been affected. The impact on society of 'our favourite drug' is ignored at our own peril.

A test of your understanding of the interaction of alcohol and society

The Health Development Agency believes that community initiatives are the best way to tackle the impact of alcohol misuse on society (Kelly 1999). If you had to evaluate the impact of a multi-agency community initiative, perhaps involving schools, pubs, police, social services, probation and occupational health services in your area, what measures would you choose to weigh up the results of this community development?

References

Abel, E. L. and Sokol, R. J. (1991) 'A revised conservative estimate of the incidence of FAS and its economic impact', *Alcoholism: Clinical and Experimental Research*, **15**, 514–24.

Alcohol Concern (1997) *Measures for Measures. A Framework For Alcohol Policy*. London: Alcohol Concern.

Anderson, P. (1995) 'Alcohol and risk of physical harm', in H. D. Holder and G. Edwards (eds) *Alcohol and Public Policy*, pp. 82–113. Oxford: Oxford University Press.

Brewers Society (1995) *Statistical Handbook*. London: Brewers Society.

Chick, J. (1994) 'Alcohol problems in the general hospital', in G. Edwards and T. J. Peters *Alcohol and Alcohol Problems*, pp. 200–10, British Medical Bulletin 50. London: Churchill Livingstone.

Clark, W. B. and Midanik, L. (1982) 'Alcohol use and alcohol problems amongst US adults', in NIAAA Monograph *Alcohol Consumption and Related Problems* (Alcohol and Health Monograph No. 1). Rockville, MD: National Institute on Alcoholism and Alcohol Abuse.

Colhoun, H., Ben-Shlomo, Y., Dong, W., Bost, L. and Marmot, M. (1997) 'Ecological analysis of collectivity of alcohol consumption in England: Importance of average drinker', *Brit. Med. J.*, **314**, 1164–8.

Conniffe, D., McCoy, D. and Whelan, K. (1996) *Alcohol and Consumption in Ireland: The Economic Factors*. Dublin: Economic and Social Research Institute.

Edwards, G., Anderson, P., Babor, T. F., Casswell, S., Ferrence, R., Giesbrecht, N., Godfrey, C., Holder, H. D., Lemmens, P., Makela, K., Midanik, L. T., Norstrom, T., Österberg, E., Romelsjo, A., Room, R., Simpura, J. and Skog, O. J. (1994) *Alcohol Policy and the Public Good*. Oxford: Oxford University Press.

Edwards, G., Marshall, E. J. and Cook, C. C. H. (1997) *The Treatment of Drinking Problems*, 3rd edn. Cambridge: Cambridge University Press.

Fombonne, E. (1998) 'Suicidal behaviour in vulnerable adolescents. Time trends and their correlates', *Brit. J. Psychiat.*, **173**, 154–9.

Foster, K., Wilmot, A. and Dobbs, J. (1990) *General Household Survey 1988*. London: Her Majesty's Stationery Office.

Glucksman, E. (1994) 'Alcohol and accidents', in G. Edwards and T. J. Peters (eds) *Alcohol and Alcohol Problems*, pp. 76–84, British Medical Bulletin 50. London: Churchill Livingstone.

Greenslade, L., Pearson, M. and Madden, M. (1995) 'A good man's fault: Alcohol and the Irish people at home and abroad', *Alcohol*, **30**, 407–14.

Kelly, M. (1999) 'Alcohol: Measuring the impact of community initiatives', *Healthlines*, **60**, 8–9.

Leon, D., Chenet, L., Shkolnikov, V. M., Zakharov, S., Shapiro, J., Rahhmanova, G., Vassin, S. and McKee, M. (1997) 'Huge variation in Russian mortality rates 1984–1994: Artefact, alcohol or what?' *Lancet*, **350**, 383–8.

Office of National Statistics (1997) *Living in Britain: Preliminary Results from the 1996 General Household Survey*. London: Office of National Statistics.

Plant, M. (1997) 'Trends in alcohol and illicit drug-related diseases', in J. Charlton and M. Murphy (eds) *The Health of Adult Britain*, pp. 114–27. London: Office for National Statistics.

Raymond, N. C., Coleman, E., Ohlerking, F., Christenson, G. A. and Miner, M. (1999) 'Psychiatric comorbidity in pedophilic sex offenders', *Amer. J. Psychiat.*, **156**, 786–8.

Royal College of Psychiatrists (1986) *Alcohol: Our Favourite Drug*. London: Tavistock.

Skog, O. J. (1980) 'Is alcohol consumption lognormally distributed?' *Brit. J. Addiction*, **75**, 169–73.

Steinglass, P., Bennett, L. A., Wolin, S. J. and Reiss, D. (1987) *The Alcoholic Family*. London: Hutchinson.

Stevenson, R. J., Lind, B. and Weatherburn, D. (1999) 'The relationship between alcohol sales and assault in New South Wales, Australia', *Addiction*, **94**, 397–410.

Thom, B. (1993) 'Social factors and alcohol abuse', in D. Bhugra and J. Leff (eds) *Principles of Social Psychiatry*, pp. 162–79. Oxford: Blackwell Scientific.

Thompson, A. (1999) 'Danger zone', *Community Care*, 22 July, 24–25.

Walberg, P., McKee, M., Shkolnikov, V., Chenet, L. and Leon, D. A. (1998) 'Economic change, crime and mortality crisis in Russia: Regional analysis', *Brit. Med. J.*, **317**, 312–18.

Zeitlin, H. (1994) 'Children with alcohol misusing parents', in G. Edwards and T. J. Peters (eds) *Alcohol and Alcohol Problems*, pp. 139–51, British Medical Bulletin 50. London: Churchill Livingstone.

Chapter 8

Smoking and the lung

Teresa D. Tetley

Key points

- *The route of delivery of a drug is important in predicting its acute effects and its pattern of use.*
- *Using the lung to absorb nicotine during smoking is associated with specific forms of damage, especially when the body's defence mechanisms are overwhelmed by frequent, high-dose use of this drug.*

Introduction

The lungs are the body's first line of defence against the toxic effects of cigarette smoke. Their complex structure is designed to remove organic and inorganic particulate material and to metabolize and 'filter out' unwanted gases and soluble material. Cigarette smoke is a complex mixture of more than 4,500 components, over forty of which are known carcinogens, with something in the order of 10^{17} free oxygen radicals per puff. Cigarette smoke also consists of a gas and particulate phase, which provides a significant challenge to the lung's defence system (Byrd 1992).

Whenever an individual inhales cigarette smoke, they essentially 'deliver' this powerful package of compounds to, first, the lung and, second, the vascular system. It is an excellent, non-invasive route for fast, effective drug delivery. It is interesting that there is marked between-subject variability in the way that they smoke (e.g. frequency and depth of inhalation) and when and how often they smoke (e.g. hourly, half-hourly). These factors are obviously significant in terms of dose effect and response, which are important physiologically (i.e. nicotine dependence) and pathologically. Deep inhalation delivers the smoke/nicotine to the whole of the respiratory tract, enhancing absorption into the blood and, therefore, maximizing the systemic metabolic effects of cigarette smoke. However, this also means that the entire respiratory tract is susceptible to its toxic effects. Shallow inhalation mostly exposes conducting airways. Compared with

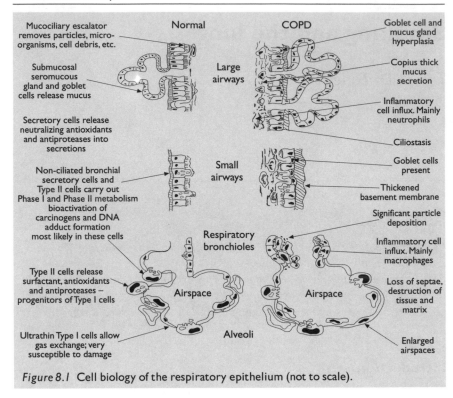

Figure 8.1 Cell biology of the respiratory epithelium (not to scale).

smoking only part of each cigarette (e.g. half), smoking to the tip of the filter or smoking high tar, filterless cigarettes, carries a higher risk of lung disease. These differences in cigarette preferences and smoking habits inevitably have a bearing on lung loading and pathological outcome that is not always easy measure.

Lung defence mechanisms

The conducting large, central and small airways (bronchi, bronchioles and respiratory bronchioli) remove deposited particles and micro-organisms via the mucociliary escalator. Thus, the epithelium (Figure 8.1) consists of ciliated and secretory cells, with bronchial glands embedded in the mucosal wall which open into the airway lumen. The bronchial gland and airway secretory cells release mucus, which consists of a very liquid phase immediately above the epithelial cells, overlaid by a thick, dense phase. The liquid phase enables the cilia, which beat in synchrony, to move the mucus along the airway towards the throat where, in healthy people, it is swallowed almost imperceptibly. The tacky consistency of the dense phase enhances trapping and immobilization of inhaled particles and micro-organisms which are cleared from the airways via the mucus 'escalator'.

Oxygenated air finally passes through the respiratory bronchioli which open into the peripheral alveolar sacs, where gaseous exchange takes place. In this region, the lung is lubricated by pulmonary surfactant, a phospho-lipid-enriched lipoprotein secreted by the alveolar Type II cells. This surface active material prevents alveolar collapse during expiration. Type II cells are the precursors to the very thin alveolar epithelial Type I cells which, although outnumbered 2 : 1 by Type II cells, are so thin that they cover 95 per cent of the alveolar surface; their basement membrane is fused with that of the underlying capillary endothelium, enabling efficient gas exchange. However, these delicate structures are very susceptible to physical and chemical damage. They are protected from micro-organisms and par-ticulates by phagocytic cells, alveolar macrophages (over 95 per cent of the total in healthy lungs) and polymorphonuclear neutrophils (PMNs), which are only a small proportion of the total but which are very important (e.g. during infection) and are sequestered during times of stress. Particles are phagocytosed and destroyed internally by 'scavenging' macrophages, which then migrate from the lung, with undigested particles in tow, via either the lymphatic system or the mucociliary escalator. PMNs release potent extra-cellular enzymes and oxidants that kill, neutralize and degrade micro-organisms and other unwanted material.

The respiratory 'secretions' (many components transude from the vascu-lar and interstitial compartments) contain numerous protective agents, including antioxidants such as non-enzymatic uric acid, ascorbic acid and glutathione and enzymes such as superoxide dismutases. They also contain antiproteases, such as alpha 1-antitrypsin and secretory leukopro-tease inhibitor, which are especially important in controlling neutrophil proteases. Secretory epithelial cells synthesize and secrete many of the anti-oxidants and antiproteases and contribute to the protective barrier in the lung. In addition, the airway epithelium is enriched by non-ciliated secretory cells (sometimes called Clara cells) which, along with alveolar Type II cells, contain the enzymes involved in xenobiotic metabolism.

Pathological effects of cigarette smoke on the lung

The two main types of lung disease that are associated with smoking tobacco are cancer and chronic obstructive pulmonary disease (COPD). The mechanisms involved are different and will therefore be discussed separately.

Chronic obstructive pulmonary disease, bronchitis and emphysema

COPD occurs in approximately 40 per cent of smokers. It results in approxi-mately 30,000 deaths per year in the UK. Cigarette smoking accounts for about 80 per cent of all COPD deaths. COPD is a general term for three conditions: emphysema; small airways inflammation and fibrosis (i.e. bronchiolitis); and mucus gland hyperplasia (i.e. bronchitis). They are all common diseases of smokers and COPD is far more common than lung cancer. Emphysema and chronic bronchitis are distinct processes which can occur independently of each other but which also often co-exist. In both conditions, a central feature in the pathology is the inflammatory response to cigarette smoke (Senior and Anthonisen 1998).

Bronchitis

The airways normally contain relatively few phagocytic inflammatory cells, with about twice as many macrophages as PMN. However, induced or expectorated sputum from bronchitics and COPD subjects shows that the this ratio reverses, becoming a one-to-one ratio in those with bronchitis but not with COPD (i.e. no small airways collapse), while in severe COPD the PMN outnumber macrophages by two to one. The total number of inflammatory airway cells also increases approximately fivefold. The combination of increased PMN, release of PMN oxidants and proteases, especially neutrophil elastase, and continued cigarette smoke exposure potentiates the inflammation and mucus hypersecretion.

A hallmark of chronic bronchitis is goblet cell hyperplasia and mucous gland enlargement leading to increased mucus release (Senior and Anthonisen 1998; Jeffery 1998). Cigarette smoke has been shown to induce this response in animal models. This is probably a defence mechanism against the cytotoxic effects of cigarette smoke. As mentioned earlier, the mucus secretions contain important protective agents, including antioxidants such as uric acid, ascorbate and glutathione. Glutathione, which is synthesized and released by the epithelial cells, has been shown to be significantly increased in smokers' lung secretions, possibly in response to increased free radicals from cigarette smoke and activated inflammatory cells. The mucus component of the cells contains numerous –SH molecules within the matrix which will also act as antioxidants to protect the underlying epithelium. In addition to antioxidants, the secretions contain epithelial cell-derived antiproteases, notably secretory leukoprotease inhibitor (SLPI), as well as serum-derived alpha 1-proteinase inhibitor. These inhibitors neutralize some bacterial proteases as well as neutrophil-derived serine proteases. Unfortunately, the 'protective' mucus hypersecretion leads to chronic bronchitis and contributes to blocking of collapsed small airways in COPD. Depending on the degree of the disease and amount of small airway damage that may have occurred, bronchitis and lung function can improve when the affected individual stops smoking (Senior and Anthonisen 1998; Jeffrey 1998).

Emphysema

Anatomically, the area affected is the respiratory acinus (Jeffery 1998; Snider 1992a,b). In smokers, the alveoli and, especially, the respiratory bronchioles become laden with inflammatory cells which can increase by up to ten times normal levels. Normally, macrophages predominate, being 85–96 per cent of the total. PMN account for less than 5 per cent and lymphocytes form the remainder. The ratio of macrophages to neutrophils does not change as markedly in the peripheral respiratory units as it does in the airways; the macrophage remains the predominant cell, although bronchiolar neutrophils may rise with advanced COPD (i.e. where chronic bronchitis exists alongside emphysema). The accumulation of macrophages in the respiratory bronchioles correlates with the major site of deposition of cigarette smoke particles and likely reflects an attempt to remove particulate matter from this region. In emphysema, the walls of the respiratory bronchioles are gradually destroyed so that the airway lumen merge into each other. The alveoli can also be affected, especially during alpha 1-antitrypsin (i.e. alpha 1-proteinase inhibitor) deficiency. In addition, the elastic network is more akin to a rotten elastic band than a super-coiled spring. Reduced surface area, loss of elastic recoil and hence increased dead space inevitably leads to reduced gas exchange and breathlessness. The breathlessness usually begins during the forties and fifties and can become so bad that it is difficult to get out of bed in the morning. Eventually, the work of breathing results in heart failure and premature death.

The damage that takes place during the pathology of emphysema cannot be reversed, although cessation of smoking will reduce the rate of decline in lung function. Unfortunately, because emphysema is often associated with bronchitis and COPD, it is difficult to diagnose. Individuals ascribe their breathlessness to lack of fitness or age. By the time it is diagnosed, the damage is severe and there is little therapeutic help.

The role of phagocytes in COPD

The neutrophil is believed to be an important mediator of bronchitis, emphysema and COPD (Snider 1992a,b). It generates high quantities (1–3 pmol/cell) of potent serine proteinases, in particular neutrophil elastase. Neutrophil elastase degrades elastin and many other connective tissue components. In addition, it activates the latent form of matrix metalloproteases, which are released by macrophages and interstitial cells during migration, tissue repair and remodelling. Macrophage metalloproteinases are increased in emphysema (Senior and Anthonisen 1998; Snider 1992b). As macrophages accumulate in the respiratory bronchioles, the peripheral lung is overwhelmed by a combination of neutrophil and macrophage-derived proteases that are capable of destroying respiratory units and disrupting small airways to induce emphysema and bronchiolitis, respectively. In bronchitis, neutrophil elastase acts as a secretagogue, stimulating epithelial mucus release, and causes ciliostasis which inhibits mucus clearance. On top of all this, it is proinflammatory and can potentiate neutrophil influx by triggering pulmonary epithelial cells to release interleukin 8, a chemokine that sequesters neutrophils from the vascular bed. There is a strong correlation between neutrophil number, IL8 and reduced lung function. Neutrophil elastase also cleaves and activates another phagocyte chemoattractant complement. The increase in neutrophil number (five- to tenfold throughout the respiratory tract) and proportion (airways) in smokers' lungs is, therefore, a major contributory factor in subsequent chronic lung disease.

Antiprotease defence

However, as mentioned earlier, there are mechanisms which normally control neutrophil elastase activity (Senior and Anthonisen 1998; Snider 1992b). Alpha 1-proteinase inhibitor (54 kDa; i.e. alpha 1-antitrypsin) transudes from the vasculature and irreversibly inhibits neutrophil elastase by covalent binding. The complex exits the lung via the vasculature. Alpha 1-proteinase inhibitor can be partially inactivated by oxidants generated by activated neutrophils and present in cigarette smoke. The methionine residue at the met-ser reactive site position 358 (total length 394) acts like a switch, so that the inhibitor can be 'switched off' at times when neutrophil elastase is required (e.g. during PMN migration). Oxidative inactivation of alpha 1-proteinase inhibitor may therefore contribute to elevated neutrophil elastase activity and COPD. It has, however, proven difficult to demonstrate that this occurs, possibly because inactive inhibitor is rapidly replaced by active molecules; alternatively, oxidative inactivation of alpha 1-proteinase inhibitor may occur only in the immediate vicinity of the neutrophil and, therefore, escape detection. Nevertheless, deficiency of alpha 1-proteinase inhibitor (resulting in approximately 15 per cent of normal circulating levels) leads to early onset (between 20 and 30 years of age) emphysema in combination with cigarette smoke exposure, illustrating the significance of this inhibitor in clearing the enzyme from its site of action.

Although alpha 1-proteinase inhibitor is an effective inhibitor of 'soluble' neutrophil elastase, it cannot inhibit the enzyme when bound to a substrate such as elastin. The epithelial cells secrete SLPI, a low molecular weight (12 kDa) reversible inhibitor, which can 'pull' neutrophil elastase off an insoluble substrate, following which the enzyme can be transferred to alpha 1-proteinase inhibitor and cleared from the lung. Interestingly, epithelial cell release of this inhibitor is stimulated by neutrophil elastase and elevated in bronchitic sputum, suggesting feedback control and protective reaction to excessive neutrophil elastase (Bingle and Tetley 1996).

Little is known about the tissue inhibitors of metalloproteinases in COPD. However, an important effect of neutrophil elastase is that it proteolytically inactivates these inhibitors, alongside activation of metallo-proteinases, inevitably augmenting the pulmonary proteolytic burden and lung damage.

Other proinflammatory factors

Apart from inhibiting antiprotease activity, cigarette smoke acts rather like neutrophil elastase as it can potentiate inflammation by inducing interleukin 8 release by airway epithelial cells and activates complement. In addition to epithelial cells, macrophages also release proinflammatory cytokines (e.g. TNFα and IL1β) and chemoattractants (e.g. IL8 and MIP1α) whose release, although not necessarily increased from each cell, is elevated overall due to the large increase in the number of macrophages in the airway lumen. Consequently, there is a continuous cycle of events that potentiates chronic inflammation in the lungs of smokers which is likely to precipitate COPD.

Abstention from smoking results in reduced inflammation and a slower rate of decline in lung function and halts the progression of emphysema, although it cannot be reversed, illustrating the significance of smoke exposure to the disease process. However, emphysema and COPD does not always occur in heavy smokers. This suggests that there are inherent factors, other than alpha 1-proteinase inhibitor deficiency, which render some individuals susceptible to cigarette smoke-induced emphysema and obstructive lung disease. The factors are unknown, but may be related to the inflammatory response, reduced lung defence mechanisms or inadequate repair processes.

Cancer

Although lung cancer accounts for less smoking-related lung disease than COPD, 90 per cent of lung cancer is due to cigarette smoking (Greenblatt et al. 1994). The risk of a smoker getting lung cancer correlates with the number of 'pack years' that they have smoked (a pack year is defined as 1 year of smoking twenty cigarettes per day). Furthermore, abstention from smoking does not immediately improve the risk, which takes years to return to near normal levels. Thus, a healthy male, current smoker of twenty cigarettes per day, has an increased mortality ratio of 23.5. If he stops smoking, this decreases to 16.7 within 2 years and to 10 and 15 within ten and fifteen years, respectively (Samet 1993). Although these figures may vary from study to study, they are invariably of the same order of magni-tude. Clearly, the cancer-related changes that occur during smoke exposure have long-lasting effects.

Metabolism of precarcinogens

Apart from free radicals, many of which will be mopped up by antioxidants and other molecules in bronchial secretions, cigarette smoke contains numerous precarcinogens; it is surprising that not every heavy smoker gets lung cancer. The evidence suggests that individual variability to the response to cigarette smoke most likely reflects a balance between:

i   the degree of metabolic activation and subsequent detoxification;
ii formation and scavenging of free radicals;
iii the amount of DNA damage and repair.

Precarcinogens in cigarette smoke are numerous and include polycyclic aryl hydrocarbons (PAH) such as benzo(a)pyrene and alkaloids such as nicotine. These compounds undergo classic Phase I and II metabolism. The family of cytochrome P450 (CYP) oxidases most often mediate oxidative Phase I metabolism, generating highly reactive metabolites, while Phase II enzymes, such as glutathione-S-transferase and N-acetyltransferase, are involved in conjugation and, ideally, inactivation of the Phase I products.

Many of these enzymes are polymorphic, and it is suggested that these genetic polymorphisms and differential enzyme reactivity contribute to individual cancer susceptibility.

Cell-specific xenobiotic metabolism

Cellular localization and airway distribution of the enzymes appears to be directly related to the toxicity and carcinogenic site; for example, compared with alveolar Type II cells, non-ciliated secretory bronchial epithelial cells preferentially metabolize benzo(a)pyrene with cytochrome P450 oxidases and metabolize over twice as much of the precarcinogen as do Type II cells. This is believed to render the bronchial cells more susceptible to adduct and tumour formation. Similarly, in mice, metabolism of naphthalene by cytochrome P450 oxidase has been demonstrated to be very variable between proximal and distal airways. It is preferentially metabolized in the distal, bronchiolar region which is where the relevant cytochrome P450 (CYP 2F2) is localized and where most naphthalene toxicity occurs, reflecting metabolism to a more toxic compound. In the same study, the Phase II enzymes were found to be evenly distributed throughout the airways.

This suggests that cell-specific metabolism of naphthalene by cytochrome P450 oxidase dictates the cellular target and initial site of tumour formation.

Metabolic activation resulting in DNA adduct formation

Amongst the many carcinogenic agents in cigarette smoke, polycyclic hydrocarbons are strongly implicated in the development of lung cancer. There are between 20 and 40 ng of benzo(a)pyrene in the particulate tar component of each cigarette smoked, and its metabolism has been studied extensively (Denissenko et al. 1996). During Phase I metabolism, the major metabolites following CYP1A1 oxidation of benzo(a)pyrene are hydro- and dihydrodiols, mainly benzo(a)pyrene-7,8-diol. Secondary metabolism of benzo(a)pyrene-7,8-diol by epoxide hydrolase (i.e. CYP) leads to the formation of the ultimate carcinogenic metabolite, benzo(a)pyrene diol epoxide, BPDE ([+/−]-anti-7b,8a-dihydroxy-9a,10a-epoxy-7,8,9,10-tetrahydrobenzo(a)pyrene). Exposure of lung epithelial cells to BPDE results in binding to DNA, predominantly forming covalent *trans* adducts at the N2 position of guanine (Denissenko et al. 1996). Importantly, CYP1A1 can be upregulated in lung epithelial cells exposed to cigarette smoke and has been shown to be significantly elevated in smokers' lung tissue. This is particularly prevalent in tissue surrounding lung tumours.

Other metabolites of components of cigarette smoke are capable of DNA adduct formation, although they may be quantitatively less important; for example, tobacco-specific nitrosamines are derived from the addictive component of cigarette smoke, nicotine, and other tobacco alkaloids. N-Nitrosamino ketone (NNK; 4-(methylnitrosoamino)-1-(3-pyridyl)-1-butanone) has been shown to be organospecific for the lung, inducing lung tumours in experimental animals, and, like BPDE, forms adducts with guanine (Richter et al. 1992; Raunio et al. 1995).

Thus, once generated, the highly reactive metabolites form adducts with DNA and are believed to be crucial in cigarette smoke-induced lung cancer.

P-32 post-labelling (which labels DNA adduct sites) from heart, lung, bronchus, kidney, aorta, bladder and oesophageal tissue from smokers showed intense labelling of heart and lung tissue; bronchial tissue was also well labelled, compared with the other tissues. Furthermore, the pulmonary P-32 post labelling intensity increased with pack years of smoking (Phillips et al. 1988). The formation of adducts obviously renders DNA susceptible to significant, irreversible mutation, the prevention of which will depend on the number and site of DNA adducts and the ability to repair affected DNA.

Enzyme polymorphisms and lung cancer

A major contributing factor to the number of DNA adducts will be the rate of metabolic transformation of precarcinogens to active carcinogenic metabolites. This can vary considerably between individuals (Raunio et al. 1995; West et al. 1997). The enzymes involved in Phase I and II metabolism are numerous, and recent studies show that, in addition, they exhibit significant polymorphism. Consequently, some enzyme polymorphisms are fast, efficient metabolizers whilst others are slow or may be completely absent (null). Amongst the family of cytochrome P450 oxidases, CYP1A1 (also often called aryl hydrocarbon hydroxylase), CYP1A2, CYP2A6, CYP2D6 and CYP2E1 are polymorphic and CYP1A1, CYP2D6 and CYP2E1 have been associated with lung cancer. Aryl hydrocarbon hydroxylase (i.e. CYP1A1) activity is inducible, and subjects with high inducibility phenotypes have a greater risk of lung cancer. In fact, the human aromatic hydrocarbon receptor complex has recently been cloned and evidence suggests that variable CYP1A1 inducibility is more likely to be related to differences in the receptor. Similarly, there appears to be a relationship between CYP2D6 metabolism of debrisoquine 4-hydroxylase, a hypotensive drug, and the incidence of lung cancer. Extensive debrisoquine metabolism due to mutation of CYP2D6 is significantly associated with lung cancer. As CYP2D6 activates the tobacco-specific precarcinogen NNK, extensive metabolism of NNK by a polymorphic CYP2D6 may contribute to cigarette smoke-induced lung cancer. This is supported by the finding that inheritance of a rare CYP2D6(C) allele, and enhanced debrisoquine metabolism, was six times more frequent in subjects with lung cancer. Defective Phase II enzymes, glutathione-S-transferase (GST) and N-acetyltransferase have also been associated with increased risk of lung cancer. The risk of lung cancer is further elevated (fortyfold) when high-risk genotypes for both CYP1A1 (i.e. extensive metabolism) and GSTM1 (i.e. null genotype) are present. This illustrates the significance of xenobiotic metabolic activation, without sufficient secondary metabolism and detoxification, in the aetiology of lung cancer (Raunio et al. 1995; West et al. 1997).

Relationship between carcinogenic metabolites and the site of DNA adducts

With increased understanding of which specific gene mutations are most commonly associated with lung cancer, it is possible to relate carcinogens to the site of adduct formation within a known gene. Mutation of the p53 tumour suppressor gene is common in many forms of cancer. However, the nature of mutations due to smoking-related cancer is often different to other mutagens; for example, certain lung cancer mutational 'hotspots' in the p53 gene have been shown, in human bronchial epithelial cells in vitro, to result preferentially from benzo(a)pyrene metabolism to BPDE and DNA adduct formation (Denissenko et al. 1996). Codon 157 in the p53 gene has been found to be one of the mutational hotspots for lung cancer, although other, less lung-specific hotspots exist. Significantly, the codon 157 mutations involve G to T (purine to pyrimidine) transversions, as might be expected following guanine adduct formation. BPDE hotspots were found on the non-transcribed DNA strand, which would be expected to be repaired inefficiently, and fits with the concept that most G to T mutations are the result of strand-bias repair with guanine on the nontranscribed strand (7,9). Other lung-tumour mutational hotspots do not involve G to T transversions (rather, A to G transitions), and are more likely the result of other mutagenic components of cigarette smoke or other mutagens (e.g. radon: Greenblatt et al. 1994; Denissenko et al. 1996).

The p53 gene normally negatively regulates cell growth and division (Greenblatt et al. 1994; Johnson and Kelley 1993; Hasday and McCrea 1992). Its mutations are most commonly observed in *small cell lung cancer* (~70 per cent) and *non-small cell lung cancer* (~47 per cent), both of which are smoking related, whereas p53 mutations are seen less often in *adenocarcinomas* (~33 per cent), which do not seem to be related to smoking. The p53 gene is described as a 'recessive oncogene' as it requires mutation of both alleles to prevent the regulatory action of p53 on cell growth and division. Even if p53 or another recessive oncogene is inactivated, it is likely that other oncogenes need to be mutated before neoplastic transformation takes place. Cigarette smoke-induced DNA adducts and subsequent mutation are very likely to be responsible for mutation of normal genes, termed 'protoncogenes', to oncogenes (Hasday and McCrea 1992).

For example, the k-ras 'dominant oncogene' (a dominant oncogene needs to be mutated at only one allele to exert oncogenic activity) is found to be mutated more often in lung adenocarcinomas from smokers than non-smokers. Furthermore, the presence of a k-ras mutation in non-small cell lung cancer has a poorer prognosis following treatment for lung cancer. κ-Ras codes for a cytosolic protein that is involved in signal transduction; mutations of these 'cytosolic' oncogenes seriously disrupt regulation of cellular activity by external factors. Nuclear oncogenes are another class of dominant oncogenes (e.g. myc and myb), which code for factors that bind to DNA and regulate DNA transcription. They usually stimulate cell proliferation, although, unlike the cytosolic oncogenes, they do not affect cell response to external stimuli. They have been shown to be present in a range of human lung cancer specimens. A series of recessive and dominant oncogenes are under investigation to establish their contribution to smoking-related lung cancer (Johnson and Kelley 1993; Hadsay and McCrea 1992).

The evidence suggests that more than one type of gene (e.g. tumour suppressor, dominant oncogene) will be associated with neoplastic transformation.

Conclusion

In conclusion, cigarette smoke causes significant lung disease and premature death, although it is remarkable, considering its known toxicological and carcinogenic potential, that many heavy smokers escape the effects. This strongly suggests that there is inherited and acquired susceptibility to cigarette smoke-induced lung disease, which is supported by the observation of common Phase I and II metabolic enzyme polymorphisms in lung tumour patients and the relationship between alpha 1-proteinase inhibitor deficiency and emphysema. The search for a relationship between inherited susceptibility factors and development of lung disease in cigarette smokers continues, possibly in the hope that identification of susceptible genotypes will convince those at risk that they should stop smoking. However, for many, the addiction to nicotine is likely to be a more powerful force that will need to be conquered by an alternative strategy to that of reason.

References

Bingle, L. and Tetley, T. D. (1996) 'Secretory leukoprotease inhibitor – partnering alpha-1 proteinase inhibitor to combat pulmonary inflammation', *Thorax*, **51**, 1273–4.

Byrd, J. C. (1992) 'Environmental tobacco smoke. Medical and legal issues', *Med. Clin. N. Amer.*, **76**, 377–98.

Denissenko, M. F., Pao, A., Tang, M-S. and Pfeifer, G. P. (1996) 'Preferential formation of benzo(a)pyrene adducts at lung cancer mutational hotspots in p53', *Science*, **274**, 430–2.

Greenblatt, M. S., Bennett, W. P., Hollstein, M. and Harris, C. C. (1994) 'Mutations in the p53 suppressor gene – clues to cancer aetiology', *Cancer Res.*, **54**, 4855–78.

Hasday J. D. and McCrea, K. A. (1992) 'Inherited predisposition to lung cancer', *Occ. Med.*, **7**, 227–40.

Jeffery, P. K. (1998) Structural and inflammatory changes in COPD: A comparison with asthma', *Thorax*, **53**, 129–36.

Johnson, B. E. and Kelley, M. J. (1993) 'Overview of genetic and molecular events in the pathogenesis of lung cancer', *Chest*, **103**, 1S–3S.

Phillips, D. H., Hewer, A., Martin, C. N., Garner, R. C. and King, M. M. (1988) Correlation of DNA adduct levels in human lung with cigarette smoking', *Nature*, **336**, 790–2.

Raunio, H., Husgafvel-Pursiainen, K., Anttila, S., Hietanen, E., Hirvonen, A. and Pelkonen, O. (1995) 'Diagnosis of polymorphisms in carcinogen-activating and inactivating enzymes and cancer susceptibility – a review', *Gene*, **159**, 113–21.

Richter, E., Schaffler, G., Malone, A. and Schulze, J. (1992) 'Tobacco-specific nitrosamines – metabolism and biological monitoring of exposure to tobacco products', *Clin. Invest.*, **79**, 290–4.

Samet, J. M. (1993) 'The epidemiology of lung cancer', *Chest*, **103**, 20S–29S.

Senior, R. M. and Anthonisen, N. R. (1998) 'Chronic obstructive pulmonary disease', *Am. J. Respir. Crit. Care Med.*, **157**(4), S139–7.

Snider, G. L. (1992a) 'Emphysema: The first two centuries. Part I', *Am. Rev. Respir. Dis.*, **146**, 1334–44.

Snider, G. L. (1992b) 'Emphysema: The first two centuries. Part II', *Am. Rev. Respir. Dis.*, **146**, 1615–22.

West, W. L., Knight, E. M., Pradhan, S. and Hinds, T. S. (1997) 'Interpatient variability: Genetic predisposition and other genetic factors', *J. Clin. Pharmacol.*, **37**, 635–48.

Chapter 9

Smoking: addiction to nicotine

Janet T. Powell

Key points

- *Multiple factors have contributed to widespread smoking of tobacco, including the biological actions of nicotine on the brain and economic developments including the industrialization of cigarette production and the resultant tax revenues related to smoking.*
- *Except for its legal status, addiction to nicotine has many features in common with dependence on the illicit drugs.*

Historical perspectives

Smoking is an ancient custom. The Greeks and Romans smoked pipes, using hemp, belladonna, mistletoe, lavender and other leaves. The American Indians have been using tobacco, in pipes, for at least 1,000 years. This habit of tobacco smoking was reported by the fifteenth and sixteenth century explorers who visited South America, the first of whom was Christopher Columbus in 1492. Every Native American village had a tobacco field, although many plants were of poor quality. The quality of the tobacco plant was improved by the New World settlers along the east coast of the USA, in the states now known as Maryland, Delaware, Virginia and the Carolinas.

The introduction to Britain of the habit of smoking tobacco in pipes is attributed to Sir Walter Raleigh. By 1620, about 15,000 kg of tobacco was imported into England, to be smoked by the rich in pipes or cigars. About this time, Seville in Spain became the first centre of the cigar industry. It was there, in the early eighteenth century, that the first cigarettes were produced from cigar scraps, for use by those who could not afford cigars. These cigarettes were made by hand, at the rate of only four cigarettes per hour. The rapid expansion of cigarette smoking had to await the Industrial Revolution in the nineteenth century, with the introduction of machines that could produce up to 120,000 cigarettes/day.

The addictive component of cigarettes is nicotine. The addictive properties of the cigarette have been recognized and exploited by the military since the time of machined cigarettes. During the Crimean War (1854–56), soldiers were offered cigarettes and smoked them in large numbers to offset the misery and deprivations of fighting, far from home, without adequate supplies (e.g. boots, food). More recently in 1992, cigarette smokers were exploited by the Iraqi military during the Gulf War: Iraqi soldiers in the trenches may not have had adequate supplies of bullets or food, but they did have cigarettes.

Current cigarette consumption and tobacco issues

The proportion of adults in Britain who smoke rose sharply from 1920 to a peak in 1975, when 53 per cent of the population smoked. Men started smoking before women. The emancipation of women started with their working role in World War II, which also gave them the freedom to buy and smoke cigarettes. During 1975, the peak smoking year, 65 per cent of men and 42 per cent of women smoked in Britain.

The first strong evidence demonstrating the health hazards of smoking, particularly with respect to lung disease and lung cancer, came from the epidemiological studies by Sir Richard Doll and others in the mid-1950s. At this time, cigarettes were unfiltered and yielding 2.5–3.0 g nicotine and 35–40 mg tar per cigarette. The filter tip and a reduction in both the tar and nicotine content of cigarettes were introduced in response to the increasing body of evidence about the health hazards of smoking. Today, filter-tipped cigarettes containing 8–15 mg tar and 0.6–1.5 mg nicotine represent about 95 per cent of the British market. The actual yield of nicotine derived from a cigarette depends on how the cigarette is smoked. The maximum yield of nicotine increases almost linearly until the butt is approached, when there is a sharp increase in nicotine yield. In contrast, the yield of tar increases exponentially towards the butt.

Health education programmes have started to bite and today about 30 per cent of the adult population of Britain smoke. Despite this downward trend in the number of smokers, at least five very worrying issues remain:

1   up to 50 per cent of teenage girls smoke;
2   the hazards of passive smoking are only beginning to emerge;
3   the increasing use of artificial fertilizers to improve tobacco crops leads to an increase in nitrates in cigarettes, with potentially increased production of the carcinogenic nitroso derivatives from aromatic tobacco components;
4   the safety (danger?) of additives used to alter the flavour of cigarettes;
5   the falling sales of tobacco in Britain and other Western countries has

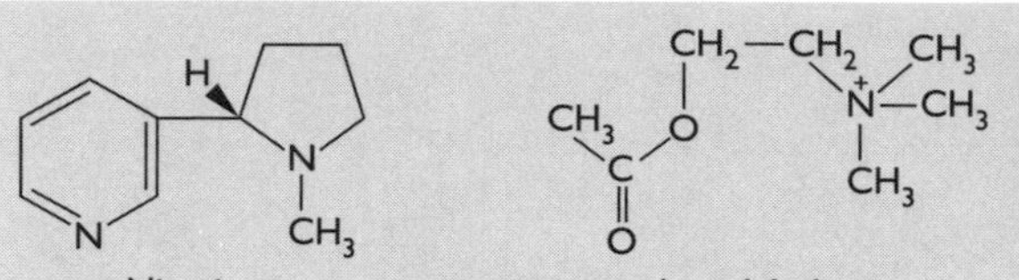

Figure 9.1 Nicotine as a structural analogue of acetylcholine.

resulted in the unscrupulous advertising and sale of cigarettes in the Third World.

The structure and metabolism of nicotine

Most of the pharmacological effects of nicotine are attributed to its structural analogy to acetylcholine (Figure 9.1). Specifically, nicotine interacts with acetylcholine receptors containing α_4 and β_2 subunits in the mesolimbic system and hippocampus.

Nicotine is both water soluble and crosses biological membranes. On smoking, there is a very high first-pass uptake of nicotine in the brain and arterial circulation: the venous concentration of nicotine is about 1/6 that of the arterial concentration. The half-life of nicotine in the circulation is ~2 hours, nicotine being oxidized by the liver microsomes principally to cotinine (Figure 9.2), with some nicotine-N-oxide also being produced. In smokers, the daytime plasma concentration of nicotine is 100–250 nM. This falls during sleeping hours to ~25 nM.

Cotinine has a much longer half-life than nicotine, ~18 hours, and provides a useful biological marker of smoking: plasma, salivary and urinary analysis of cotinine has been used for this purpose. In smokers, the plasma concentration of cotinine ranges from 200–2,000 nM. Cotinine also has pharmacological effects which are separate from those of nicotine and contribute to mood, withdrawal and effects on steroid hormone metabolism. Other metabolic products of nicotine, particularly when the tobacco leaf contains a high concentration of nitrates, include nitrosonornicotine (NNN) and 4-methylnitrosamino-1-(3 pyridyl)-1-butanone (NNK) (Figure 9.2). Both NNN and NNK are powerful respiratory tract carcinogens in animals. Following their metabolic conversion to electrophiles, these nitroso-products interact with guanine residues in DNA to form 0^6 methylguanine or N^7 guanine derivatives. The carcinogenic potential of NNN, NNK and other biotransformation products generated from aromatic compounds yielded by tobacco combustion depends, at least in part, on the genetic background of the exposed individual (see Chapter 8).

The reason for this dependence on genetic background is thought to arise from the different rates amongst individuals of biotransformation of nicotine and other aromatic compounds to dangerous metabolites; for instance,

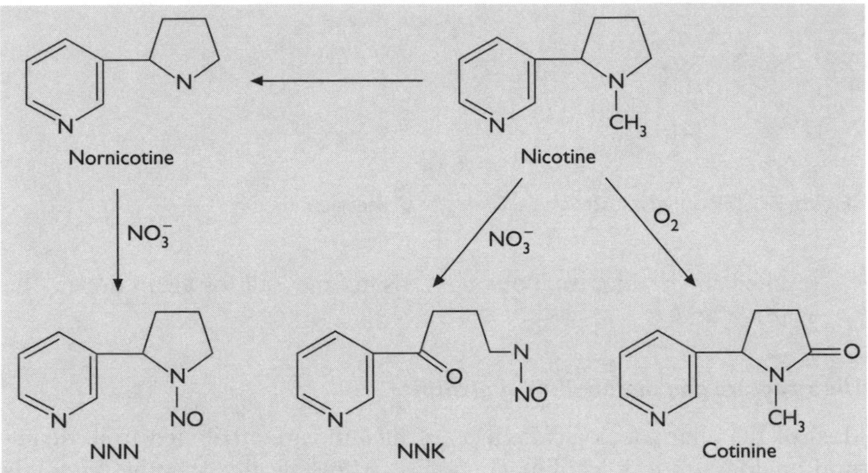

Figure 9.2 The metabolism of nicotine.

smokers who are fast acetylators of aromatic amines appear to have a reduced risk of both lung and bladder cancer than slow acetylators.

Smoking and neurotransmitter release in the central nervous system

Nicotine binds to cholinergic receptors in the nucleus accumbens to effect the release of dopamine and to modulate dopaminergic transmission through the limbic system, including the midbrain, amygdala and frontal cortex. The nucleus accumbens is a focal point for the action of many drugs of abuse, including alcohol, cocaine, heroin and amphetamines. Gateways introduced by nicotine or any of these other drugs may facilitate addiction to other addictive drugs. The release of dopamine in the nucleus accumbens may mediate pleasurable sensations and the rewarding effects of smoking. Although dopamine is the principal neurotransmitter released in the brain by smoking, noradrenaline (leading to appetite suppression), acetylcholine (leading to arousal), vasopressin (leading to memory) and endorphin (leading to anxiety reduction) also are released. Cotinine leads to the release of serotonin (5-hydroxytryptamine). On cessation of smoking, the reduced production of serotonin in the hippocampus leads to depression and withdrawal symptoms.

Nicotine has other important effects in the central nervous system. Post-mortem studies using [^{3}H]-nicotine binding have shown that smoking results in a doubling in the density of nicotinic receptors in the hippocampus gyrus rectus, the median raphe nuclei and the cerebellum.

In experimental models of nicotine abuse, usually rats, it also has been demonstrated that chronic nicotine administration leads to enhanced

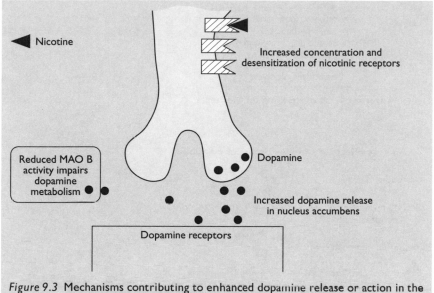

Figure 9.3 Mechanisms contributing to enhanced dopamine release or action in the central nervous system of smokers.

mesolimbic dopamine secretion, enhanced locomotor activity and desensitization of mesolimbic nicotinic receptors. These nicotinic receptors are present on the membranes of both the nerve terminal and the somatodendritic regions of this neural pathway. The somatodendritic nicotonic receptors in the ventral tegmental area of the brain are the most likely candidates for mediating these responses.

The increased release of dopamine, effected by habitual nicotine intake from smoking, could be potentiated by the effect of smoking to inactivate monoamine oxidase B, the enzyme responsible for metabolizing dopamine to dihydroxyphenylacetate in glial cells. Thus, nicotine (smoking) produces a positive-feedback system to reinforce the release and effects of dopamine (Figure 9.3). Whilst this leads to addiction in smokers, this positive feedback loop has been used therapeutically to potentiate dopamine release/action in Parkinson's and Alzheimer's disease.

Actions of nicotine in the peripheral nervous system

In the peripheral nervous system, nicotine interacts with cholinergic receptors to trigger the release of catecholamines. This has many unwanted effects on the cardiovascular system including increases in heart rate, coronary blood flow, cardiac output and fatty acid metabolism. These effects have been reviewed comprehensively by Benowitz (1996). Smoking also has numerous other effects on the cardiovascular system which act together with the increased catecholamine release to cause cardiovascular disease.

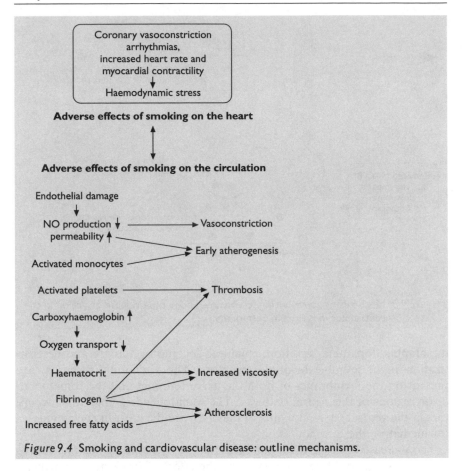

Figure 9.4 Smoking and cardiovascular disease: outline mechanisms.

These pathways are outlined in Figure 9.4. Nicotine itself is toxic to the endothelium, causing the apoptosis and shedding of endothelial cells into the circulation. Other products of tobacco combustion impair oxygen transport, endothelial function and platelet behaviour. Cotinine has complex effects on the synthesis of adrenal and steroid hormones, which contribute to the earlier menopause in female smokers and decreased fecundity of male smokers.

Hazardous components of mainstream smoke

Apart from nicotine, mainstream smoke delivers carbon monoxide, hydrogen cyanide, ammonia, acrolein, acrylamide, cadmium and a diverse array of aromatic compounds (including benzene, benzopyrene, phenol, catechol, pyridine and nitrosamines) to the lung. Many of these aromatic compounds are carcinogens or co-carcinogens. Cadmium causes renal disease and

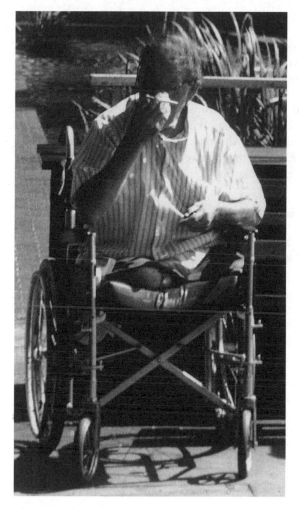

Figure 9.5 Addiction to nicotine: my cigarettes or my legs?

perhaps hypertension. Carbon monoxide and hydrogen cyanide impair oxygen transport and utilization. Those who crave to smoke should be aware of these noxious by-products.

Smoking cessation

The fairly recent usage of nicotine replacement therapy (nasal spray, chewing gum or dermal patches) to improve the success of smoking cessation programmes is other evidence to support the addictive power of cigarettes. After all, why else would a man lose both his legs from gangrene rather than give up smoking (Figure 9.5)? Smoking results in a 'rush of nicotine' to the circulation and brain. This contrasts with the slow, minimal absorption of nicotine from replacement therapies. It is likely that this difference results in the relatively modest success of nicotine replacement programmes to facilitate smoking cessation. The genetic background of an individual smoker also may contribute to both the damaging health of smoking and the ease with which the individual can give up smoking. However, like alcohol, smoking is an addiction accepted and taxed by society. In this sense, the management of 'no smoking' programmes will remain different to the management of addiction to opiates and other drugs discussed in this book.

Bibliography

Bastecchi, C. E., MacKenzie, T. D. and Schrier, R. W. (1994) 'The human costs of tobacco use', *New Engl. J. Med.*, **330**, 907–12, 975–80.

Benowitz, N. L. (1996) 'Pharmacology of nicotine: Addiction and therapeutics', *Ann. Rev. Pharmacol. Toxicol.*, **36**, 597–613.

Iversen, L. L. (1996) 'Smoking ... harmful to the brain', *Nature*, **382**, 206–7.

Stokerman, I. P. and Shoaib, M. (1991) 'The neurobiology of tobacco addiction', *Trends Pharmacol. Sci.*, **12**, 467–73.

Chapter 10

Helping people to stop smoking

Nicola Griffiths

Key points

- *The two main reasons people give for stopping smoking are health concerns and the cost of smoking.*
- *Once a person has decided or been prompted to quit smoking, a number of intervention strategies are available to them.*
- *Community based and motivational interventions aim to encourage more people to try to stop smoking.*
- *Treatment interventions aim to increase the chances of a quit attempt being successful.*
- *It is suggested that there is a case for more specialist clinics to treat motivated but dependent smokers.*
- *As the population smoking prevalence declines, so the balance of interventions needs to shift.*
- *Nicotine replacement therapy can double the chances of a quit attempt being successful.*

Introduction

For people who smoke, smoking is the single most important health risk that they can change to improve the length and quality of their life (Department of Health 1998). Yet, for the majority of smokers, it is the hardest single lifestyle change that they will ever attempt to make. Twenty-seven per cent of the UK population currently smoke (OPCS 1994) and may be faced by this dilemma.

Most smokers start smoking cigarettes before the age of 16 and the habit of smoking quickly develops. Many adult smokers continue smoking, not through unfettered choice, but because they are addicted or dependent on the nicotine in cigarettes. In a major review of nicotine addiction, the US Surgeon General concluded that 'the pharmacological and behavioral processes that determine tobacco addiction are similar to those that determine addiction to drugs such as heroin and cocaine' (US Surgeon General 1988).

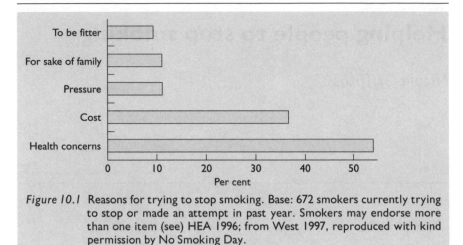

Figure 10.1 Reasons for trying to stop smoking. Base: 672 smokers currently trying to stop or made an attempt in past year. Smokers may endorse more than one item (see) HEA 1996; from West 1997, reproduced with kind permission by No Smoking Day.

Thus, for both pharmacological and psychological reasons, most smokers experience difficulty in quitting the habit.

Reasons for stopping smoking

The Health Education Authority (HEA) periodically carries out tracking surveys to assess smoking patterns in England. The most recent of these surveyed 1,911 smokers in April and May 1996 (HEA 1996). The smokers were asked about their experiences of trying to stop smoking. Results show that at any one time, 54 per cent of smokers want to stop smoking and 16 per cent of smokers are actually trying to do so. Thirty-one per cent have tried one or more times to stop smoking in the past year. Figure 10.1 highlights the main reasons people give for wanting to quit.

Clearly, the two main reasons for quitting are health concerns and the cost of smoking. Therefore, it is not surprising that the health benefits of quitting and 'what you can spend the money on if you quit' are two of the strategies often used to help boost smokers' motivation to quit. Motivation to quit has been shown to have a dramatic effect on success rates. If a person's motivation to quit smoking is high, they will have a greater chance of quitting, with or without the help of products.

Scope for cessation

Evidence that smoking interventions are effective in terms of prevention, cessation and cost-effectiveness is essential for identifying key targets and priorities. In terms of prevention, the NHS Centre for Reviews and Dissemination (1999) has shown that the evidence supports a *mix* of different methods (such as school-based or community interventions) with proper co-ordination to ensure that a *range* of young people can be engaged. Likewise, a comprehensive service for smoking cessation needs a co-ordinated mix of interventions with appropriately trained staff (Hodgson 1998). Stopping smoking is, however, a personal decision which smokers have to carry through themselves. One-half of smokers who quit do so by simply deciding to stop (HEA 1996), but they may be helped by the informal support of family, friends or colleagues. The rest attempt to quit using one or more of the methods presented below.

Population-based interventions

Community or population-based smoking cessation projects generally aim to increase the number of smokers within the community who make an attempt to stop smoking, and to provide a net reduction in smoking prevalence within the target community. Such interventions aim to create a social climate that does not support tobacco use, by intervening through social structures within a community (i.e. radio advertising). The advantage of this public health approach is being able to reach a sizable proportion of the population; however, this approach does have the lowest efficacy rate (Prochaska 1996). Furthermore, it seems that in countries or communities in which the health education message is widely accepted, such as the UK, further population interventions produce rather small effects and may have absolutely no influence at all on heavy (highly dependent) smokers.

Brief smoking cessation interventions by health professionals

Study days designed to train health professionals to provide smoking cessation interventions are effective in increasing the number of occasions prompts and advice to quit are given (Silagy *et al.* 1996a). There is a modest, but still statistically significant, effect on the outcome of cessation. However, health professionals do need regular prompts, reminders and training in order to retain effectiveness.

Brief smoking interventions which may raise the issue of smoking and suggest methods to quit during routine consultations are a particularly cost-effective method of increasing the number of smokers who stop smoking, and beneficial to motivated smokers who wish to stop. The British population rate advice from their general practitioner as the most trustworthy source available (Budd 1991). One study (Sanders *et al.* 1989) suggests 5 per cent of smokers will stop by simply being told to do so by their GP. Whilst this change seems small, if practised consistently and persistently, it will have an important effect on the population's smoking habits and overall prevalence. Clinicians understandably fail to appreciate the population impact of their efforts given that relatively few patients quit in response to low-intensity interventions (Lichtenstein *et al.* 1996). Simple advice

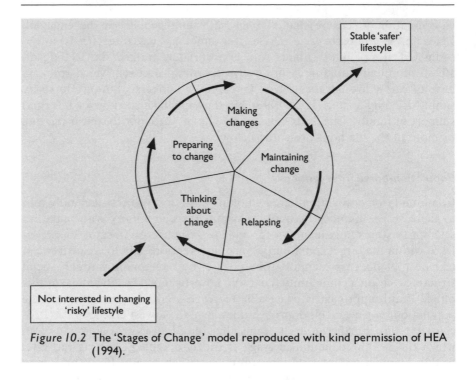

Figure 10.2 The 'Stages of Change' model reproduced with kind permission of HEA (1994).

does have an effect on population cessation rates (Silagy and Ketteridge 1996) and needs to be exploited.

Motivational interviewing

Stopping smoking is more of a process than a single event. Interventions by primary health care, and increasingly hospital clinicians and nurses, need to relate to this using methods to boost the motivation of the smokers with each key stage. One such model, based on motivational techniques, is the 'Stages of Change' model.

This model, which has been adapted by the HEA (1994), recognizes that, like many other behaviours, smoking cessation is often better described as a cyclical activity rather than a discrete stage. It is suggested that smokers typically progress through the stages given in Figure 10.2.

Relapse is often common in smokers attempting to quit, causing them to move through the cycle once more. It is reported that many smokers will go through this process seven or eight times before being successful.

On a practice or population basis, it would appear that any level of activity which promotes smoking cessation in primary care, from minimal intervention upwards, will have an effect, and that success is proportional to the level of input by professionals rather than to its specific content (Lennox

1992). However, for some people who wish to quit, this approach is clearly not enough.

Specialized smoking cessation clinics and treatments

Public health campaigns and brief interventions, when appropriately targeted, are cost-effective methods of reducing the number of smokers and producing a substantial health gain. Consistent use of these measures will leave a large number of smokers who have tried to quit on their own, and following advice from their GP, without success. This select group may respond to a specialized smokers clinic. The success rate of clinics ranges from 16 per cent (Foulds 1993) to 27 per cent when Nicotine Replacement Therapy (NRT) is provided (Silagy *et al.* 1996a). The advantages of such clinics are the availability of NRT products and providing a level of support necessary to ensure proper usage. However, whilst clinics have the highest efficacy with better success rates, they have the lowest reach (Prochaska 1996), probably due to the limited number of clinics in the UK at this time.

A typical clinic would involve the smoker attending for an initial assessment and preparation, then on five occasions 1 week apart. The first treatment session would be when the smoker stopped. The main element of each session would consist of a group discussion in which smokers provide mutual support and encouragement.

Nicotine replacement products

There are two main groups of smoking cessation aids: those containing nicotine and those which use other substances/forms of treatment. NRT (using nicotine gum, transdermal patch, nasal spray, inhaler) is designed to help break the habit of smoking and take the edge off withdrawal symptoms, while continuing to provide a reduced dose of nicotine. Withdrawal from the *drug* is tackled when the habit of *smoking* has been broken.

Overall, all four forms of NRT are significantly more effective than placebo or no NRT in helping smokers achieve abstinence (Silagy *et al.* 1996b). There is no statistical difference in the effectiveness of the four types of NRT. When pooled, the results show 18 per cent are abstinent after 12 months. Abstinence rates are higher in volunteers from a community group or setting and in smoking cessation clinics, as these clients tend to be highly motivated to quit and the setting is controlled. This does however, represent a 77 per cent increase in the odds of abstinence with the use of NRT. Recent studies of treatment outcomes (Bolliger 2000) find that NRT not only improves the odds of stopping smoking, it also reduces the number of cigarettes consumed by the patients who are unable to stop altogether. Details of the four types of NRT are given in Table 10.1.

Table 10.1 Summary of nicotine replacement products (NRT)

Product	Dosage	How it works	Main side effects
Gum	2 mg and 4 mg. 15 pieces a day for 3 months. Chew each piece slowly for 30 minutes. Available at local pharmacies (no prescription necessary)	Nicotine is slowly absorbed through the buccal mucosa in the mouth	Mild jaw ache, indigestion and irritation of the mouth and throat
Patches	16 hour and 24 hour patches. A new patch is applied each day for 3 months. Available at local pharmacies (no prescription necessary)	Nicotine is slowly absorbed through the skin into the body over the course of the day/ night	Disturbed sleep (24 hour patch), early morning cravings (16 hour patch) and reddening of the skin where patch is applied
Spray	One spray to one or both nostrils hourly. One applicator lasts approximately 16 days. Use for 3 months. Available on private prescription only	Nicotine absorbed rapidly through nasal membranes	Irritation of the nose and throat which can be very unpleasant at first
Inhaler	Inhaled for 10 minutes every hour. Six capsules per day. Still in trials in the UK	Nicotine vapour is inhaled from a nicotine-impregnated plug through a type of cigarette holder	Mouth and throat irritation and coughing

Non-nicotine products

Other cessation aids are available from pharmacists without a prescription including lozenges, capsules and tablets. Some contain small doses of nicotine and aim to reduce the cravings whilst others contain silver acetate which produces an unpleasant taste when smoking. Few of these products have been clinically tested although they may be of assistance to some people. It should be noted that many of these products do not give clear instructions or a recommended dose.

Dummy cigarettes consist of a plastic tube containing a variety of substances such as menthol or eucalyptus, and may help those smokers who have problems overcoming the ritual of handling cigarettes. There is, however, little more than anecdotal evidence that these are effective. There is also little evidence to support the effectiveness of either acupuncture or hypnosis as a means of eliminating smoking behaviour, but such methods may suit some smokers who want to try this form of treatment. Herbal cigarettes are not recommended as an aid to giving up smoking because they produce both tar and carbon monoxide. Some brands have a tar content equivalent to tobacco cigarettes. In addition, the use of herbal cigarettes reinforces the habit of smoking, which smokers need to overcome.

Summary of implications

It is the role of the health organizations to implement strategies to reduce the prevalence of smoking by ensuring individuals who smoke are aware of the risks they are taking with their health. If they wish to quit, they should have access to the necessary support to do so. This support needs to range from the availability of one-to-one advice by the primary health care team, to nicotine-replacement products, and, for heavily dependent smokers, access to specialized smoking cessation services such as clinics. An integration of services is needed to meet the needs of all smokers who want to quit. As population smoking prevalence declines, so the balance of interventions should shift from motivational to treatment approaches. Currently, it is suggested that 22 per cent of smokers (Di Clemente *et al.* 1991) need treatments such as NRT and brief support, whilst 5 per cent (Prochaska *et al.* 1993) need NRT and intensive support. This suggests that there is a case for more specialist clinics to treat motivated but addicted smokers, and train health professionals to apply effective smoking cessation methods as part of their routine work.

Smokers who do not have access to services might wish to construct their own 'Do It Yourself' programme (West 1997), based around the ingredients of the most successful clinics, to improve their own chances of success.

A 'DIY' stop smoking programme

1   *Make the commitment:* set a date and time for giving up and stick to it; go for complete abstinence; do not allow any slips; put stopping at the top of your priority list.
2   *Make the break:* dispose of all smoking materials and paraphernalia and avoid situations where you will be tempted to smoke.
3   *Do it with others:* find others who would like to give up and agree to support each other and keep in daily contact; make a pact not to let each other down.
4   *Ease the withdrawal symptoms:* use NRT and follow the manufacturers' instructions ensuring that you use enough, for long enough.

Reproduced with kind permission of West (1997).

References

Budd, J. and McCron, R. (1991) *Communication and Health Education*. London: Health Education Council.

Bolliger, C. T. (2000) 'Practical experiences in smoking reduction and cessation', *Addiction*, **95**(Suppl. 1), S19–S24.

Department of Health (1998) *Smoking Kills*. London: DoH.

Di Clemente, C. C., Prochaska, J. O., Fairhurst, S. K., Velicier, W. F., Velasquez, M. M. and Rossi, J. S. (1991) 'The process of smoking cessation: An analysis of pre-contemplation and preparation stages of change', *J. Consul. Clin. Psychol.*, **59**, 294–304.

Foulds, J. (1993) 'Does nicotine therapy work?' *Addiction*, **88**, 1473–8.

Health Education Authority (1994) 'Helping people change: Health promotion in primary care', *Trainers Manual*. London: HEA.

Health Education Authority (1996) 'National Adult Smoking Campaign Tracking Survey (Wave 3)', unpublished, London: HEA.

Hodgson, P. (1998) 'Smoking interventions covering every angle', *Healthlines*, **50**, 18–19.

Lennox, A. S. (1992) 'Determinants of outcome in smoking cessation', *Brit. J. Gen. Practice*, **42**, 247–52.

Lichtenstein, E., Hollis, J. R., Severson, H. H., Stevens, V. J., Vogt, T. M., Glasgow, R. E. and Andrews, J. A. (1996) 'Tobacco cessation interventions in health care settings: Rationale, model, outcomes', *Addictive Behaviors*, **21**, 709–20.

NHS Centre for Reviews and Dissemination (1999) 'Preventing the uptake of smoking in young people', *Effective Health Care*, **5**.

OPCS (1994) *General Household Survey: Living in Britain*. London: Office of Population Censuses and Surveys.

Prochaska, J. O. (1996) 'A stage paradigm for integrating clinical and public health approaches to smoking cessation', *Addictive Behaviors*, **21**, 721–32.

Prochaska, J. O., Di Clemente, C. C., Velicier, W. and Rossi, J. S. (1993) 'Standardized, individualized, interactive and personalized self-help programmes for smoking cessation', *Health Psychol.*, **12**, 394–405.

Sanders, D., Fowler, G., Mant, D., Fuller, A., Jones, L. and Marziller, J. (1989) 'Randomized control trial of anti-smoking advice by nurses in general practice', *J. Roy. Coll. Gen. Practitioners*, **39**, 273–6.

Silagy, C., Mant, D., Fowler, H. G. and Lodge, M. (1992) 'Meta-analysis on the efficacy of nicotine replacement therapies in smoking cessation', *Lancet*, **340**, 342–9.

Silagy, C. and Ketteridge, S. (1996) 'The effectiveness of physician advice to aid smoking cessation', *The Cochrane Library*, **2**, 21 May 1996.

Silagy, C., Lancaster, T., Fowler, G. and Spiers, I. (1996a) 'Effectiveness of training health care professionals to provide smoking cessation interventions: Systematic review of randomized controlled trials', *The Cochrane Library*, **2**, 2 June 1996.

Silagy, C., Mant, D., Fowler, G. and Lancaster, T. (1996b) 'The effect of nicotine replacement therapy on smoking cessation', *The Cochrane Library*, **3**, 1 May 1996.

US Surgeon General (1988) *The Health Consequences of Smoking – Nicotine Addiction*. Washington, DC: US Government Printing Office.

West, R. (1997) *Getting Serious About Stopping Smoking – A Review of Products Services and Techniques* (A report for No Smoking Day 1997). London: No Smoking Day.

Chapter 11

Molecular basis of addiction

Jackie de Belleroche

Key points

- *The mode of action of a number of drugs of abuse has been elucidated recently identifying specific molecular targets.*
- *Opiate drugs target the μ receptor which is involved in mediating the actions of the naturally occurring opioid peptides whose physiological role is concerned with nociceptive, homeostatic, neuroendocrine and limbic function.*
- *Cocaine and Ecstasy target dopamine and 5-hydroxytryptamine transporters which are important synaptic proteins involved in terminating neurotransmitter action at the synapse by rapid removal. When these transporters are inhibited, the action of the neurotransmitter (e.g. dopamine) is prolonged.*
- *The mesolimbic and mesocortical dopaminergic pathways originating in the midbrain and terminating in the forebrain play a key role in reward pathways or 'pleasure centres' which are central to the actions of most drugs of abuse.*
- *Tetrahydrocannabinoids/marijuana target a specific receptor present in limbic areas of the brain including dopaminergic regions.*
- *The PCP target is a subtype of glutamate receptor involved in specialized adaptive responses such as synaptic potentiation.*

Drugs of abuse have been used for centuries, but their mode of action has only recently come to light. We are, however, now able to pinpoint the sites of action for some of the more common drugs of abuse which mediate their addictive effects (Table 11.1).

Less well understood are the actions of drugs of abuse, such as solvents and LSD, and the complex behaviours elicited by drug taking, the development of tolerance, the processes responsible for physiological and psychological dependence (i.e. most of the addictive process). So, although we know about the first port of call and the initial reward pathway which

Table 11.1 Molecular basis of addiction

Heroin/morphine	Opiate receptor–μ receptor (δ)
Cocaine	Dopamine transporter
Tetrahydrocannabinoids	Cannabinoid receptor
Ecstasy	5-HT transporter
Phencyclidine	Glutamate receptor (NMDA receptor)

Table 11.2 The stimulants (produce a sensation of extreme mental and physical power)

Amphetamine	Benzedrine, Dexedrine and methamphetamine
Cocaine	Cocaine hydrochloride and crack cocaine (free base)

leads to acquisition of drug-taking behaviour, much of the subsequent events which maintain the behaviour are yet to be elucidated.

The focus of this chapter will be the most widely used drugs of abuse indicated in Table 11.1. Misuse of these drugs is known to kill several hundred individuals every year in the UK alone, which is particularly devastating in the case of young people in their teens or twenties, but the most far-reaching effects of these drugs relate to their effects on the quality of life, ability to sustain a job or stable relationships.

Drugs of abuse, as with tobacco and alcohol, are not just the preserve of adults but are also widely used by children. In an NHS advisory service (1995) survey, it was shown that 60 per cent of under-16-year-olds had been offered drugs and one-third of these had actually succumbed and tried drugs themselves. In an earlier study of London children, 13 per cent of 14-year-olds were shown to have taken cannabis and 4–8 per cent solvents, synthetic drugs such as LSD, amphetamines and Ecstasy.

The stimulants

The first class of drugs to be considered will be the stimulants (Table 11.2). This class of drugs will be used to define the basic process of addiction.

The first widely used stimulants were the amphetamines. Initially used in the Second World War for their properties in keeping airmen awake on long missions and later for their fairly widespread use in dieting, they are now frequently used as an adjunct to other preparations. The sought-out stimulant from the latter part of the 1980s and through the 1990s has been cocaine; one of the main reasons for its gain in popularity being the shift to a more effective mode of delivery for rapid absorption of the drug (Woolverton and Johnson 1992; Strang et al. 1993). The switch from the early days of chewing leaves of the coca plant to use of the purified salt by

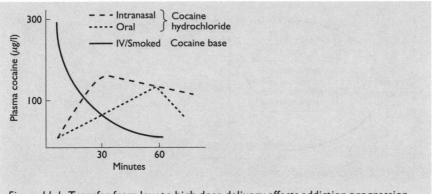

Figure 11.1 Transfer from low to high dose delivery affects addiction progression.

the nasal route progressed to the delivery in a high dose by intravenous administration or smoking (free base vaporizes on heating to produce a maximal concentration in the blood plasma at 30 sec) (Figure 11.1). Cocaine in this form is a most powerful reinforcer. Primates given access to the drug by lever pressing will self-administer continuously until they collapse exhausted. This is equivalent to the binge state in humans where readministration occurs as frequently as every 10–30 min. This is followed by 'crash' (depression, agitation, anxiety and paranoia) and a dysphoric syndrome in which everything appears bleak and pointless (Gawin 1991).

Molecular basis of addiction

The effects of cocaine when first taken give a profound subjective feeling of well-being and alertness, increased physical strength and intensification of feelings of pleasure (sexual and emotional) and self-confidence. This represents the activation of a reward pathway (Figure 11.2) giving rise to the feeling of euphoria which subsequently causes the individual to repeat the process by continually seeking a renewal of that experience by continued drug taking. However, with drugs of abuse the same experience cannot be repeated with the same dose of the drug, and increased doses are required to produce the same effect, known as tolerance. With continued drug taking, a dependence is developed which is particularly evident from the unpleasant side effects which result on abstinence from the drug (withdrawal signs). This model of action appears to be universal (Figure 11.3) and will be used as a basis for discussing drug action in this chapter. Looking at the mode of drug action in two parts, there is, first, the mechanism whereby activation of the endogenous reward pathway occurs (*acquisition*) and, second, the adaptive response that occurs after chronic drug abuse (*tolerance/ dependence*).

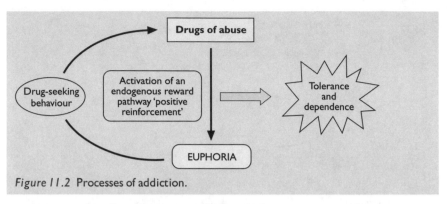

Figure 11.2 Processes of addiction.

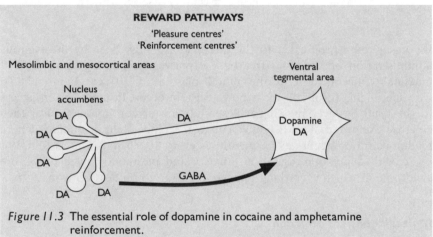

Figure 11.3 The essential role of dopamine in cocaine and amphetamine reinforcement.

Activation of endogenous reward pathways

Studies in experimental animals have pinpointed a pathway which is fundamental to the reward pathway or pleasure centre. This is known as the mesolimbic dopaminergic system (Figure 11.3). When electrodes are implanted in the region of the cell bodies which lie in the ventral tegmental area of the mid-brain, animals will press a lever to receive continued electrical stimulation, a property which is almost exclusively the preserve of this region and the sites connected with it (Koob and Bloom 1988). This process is known as intracranial self-stimulation. There is convincing evidence to show that this pathway is essential to manifest the addictive properties of both cocaine and the opiates (Kuhar *et al.* 1991; Gawin 1991). If this dopaminergic nucleus is selectively destroyed by 6-hydroxydopamine, or if dopamine receptor antagonists (D_2) are used, then cocaine self-administration is abolished. Increased dopamine release in the target tissue, the nucleus

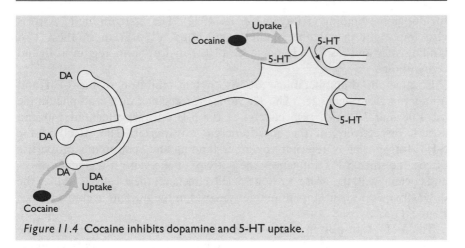

Figure 11.4 Cocaine inhibits dopamine and 5-HT uptake.

accumbens, when cocaine is applied, can be demonstrated by microdialysis (a method which allows superfusion of a localized brain region). The output from the nucleus accumbens is to the ventral pallidum, frontal cortex and thalamus, and it has been known for some time that dopaminergic activation in this region produces profound motor arousal, activation and psychomotor stimulation which is consistent with the behavioural pattern seen with cocaine use and the associated enhanced dopaminergic neurotransmission.

Molecular targets of cocaine

Cocaine acts by inhibiting dopamine removal from the synapse (Figure 11.4). The rapid removal of neurotransmitter from the synapse is essential for terminating synaptic action and regulating neurotransmission. The most common mechanism for terminating synaptic action is by use of high-affinity uptake systems (transporters) which remove the neurotransmitter into surrounding nerve endings and other neuronal or glial cells. This mechanism of inactivation is used for the amino acid transmitters such as glutamate and GABA and the catecholamines and indolamines, dopamine, noradrenaline and 5-HT.

A number of these transporters have recently been cloned (Kilty *et al.* 1991; Pacholczyk *et al.* 1991; Hoffman *et al.* 1991; Uhl 1992) and their structure elucidated. They form part of a structurally similar 'super' family of proteins of approximately 600 amino acids in length with twelve hydrophobic regions which represent transmembrane domains. Both the C- and the N-terminal domains are intracellular. The transmembrane domains are important in ligand binding, and it has been shown by site-directed mutagenesis of the dopamine transporter that ^{79}Asp and ^{356}Ser, which lie in the first and seventh transmembrane domain, respectively, are essential for dopamine binding (Kitayama *et al.* 1992). The first residue (^{79}Asp) is also important in cocaine binding, indicating an overlap in these two binding sites and also the importance of the hydrophobic region in ligand recognition.

Cocaine and d-amphetamine are also potent inhibitors of the 5-HT and noradrenergic transporter. The K_i values for cocaine and d-amphetamine are 1.08 and 3.18 nM, respectively, at the 5-HT transporter and 140 and 56 nM, respectively, at the noradrenergic transporter. The effect on the 5-HT transporter is relevant to understanding the reinforcing properties of cocaine since 5-HT influences the activity of the ventral tegmental area, and cocaine will, therefore, prolong 5-HT effects on these cells as well as the downstream effects at dopaminergic synapses in the nucleus accumbens (see Figure 11.3).

The 5-HT transporter is also an important target in understanding the mode of action of MDMA (3,4-methylenedioxy-methamphetamine, 'Ecstasy'). MDMA is highly potent on the 5-HT transporter with a K_i of 186 nM. Although the noradrenergic transporter is not thought to contribute to the rewarding properties of cocaine, it is very important in mediating the side effects of drug use, in particular the cardiovascular toxicity of cocaine. The effect of cocaine on blocking noradrenaline uptake will produce a sympathomimetic effect, affecting the heart and peripheral vasculature. This is responsible for the side effects of drug use which result in stroke, arrhythmia, cardiomyopathy, myocardial infarction within minutes or hours of use in some cases, coronary vasospasm and myocarditis which is evident in 20 per cent of cocaine deaths.

The euphoriant effects of cocaine are seen at plasma concentrations of 0.3–1 µM (32 mg intravenously or 2 mg/kg orally will produce this effect). These concentrations are close to those at which the drug is effective against the three transporters. The toxic effects of cocaine use are seen at only slightly elevated concentrations. Plasma concentrations of 4–25 µM cocaine have been found in cases that have been fatal.

The opiates

We have known about the actions of opium since the times of the ancient Greeks. However, despite the extensive use of morphine and its derivative diacetyl morphine (heroin) medicinally for pain relief and as a drug of abuse, it was only in the 1970s that the first endogenous 'opiates' methionine and leucine enkephalin were discovered. These are called opioid peptides (Rodgers and Cooper 1988). Some twenty or so opioid peptides have now been characterized and all contain the pentapeptide sequence of either methionine or leucine enkephalin, and are derived from one of three precursor proteins each encoded by a different gene (Table 11.3). Despite the

Table 11.3 Endogenous opioid peptides (there are more than twenty) are included in three peptide families

Gene product precursor	Physiologically active peptide (number of amino acids)	Main receptor
Preproenkephalin (PPE)	Met-enkephalin (5)	δ
Proopiomelanocortin (POMC)	β-Endorphin (31)	μ
Preprodynorphin (PPD)	Dynorphin (17)	κ

Opiate receptors defined biochemically (1973).
Endogenous ligands for the opiate receptor (identified 1975): methione enkephalin (Tyr-Gly-Gly-Phe-Met) and leucine enkephalin (Tyr-Gly-Gly-Phe-Leu).

similar analgesic properties of opioids and opioid peptides, none have the same strong euphoriant properties as morphine.

The receptors mediating the actions of these three families of opioid peptides (exemplified by β-endorphin, methionine enkephalin and dynorphin) corresponded well to the pharmacologically characterized opiate receptors μ, δ and κ, respectively. Whereas all three receptors are involved in the analgesic properties of opiates, it is the μ and δ receptors in the nucleus accumbens of the ventral tegmental area and prefrontal cortex that are most important in reward pathways (Reisine and Bell 1993; Watson and Girdlestone 1996).

Morphine reward pathways

The dopaminergic cells in the ventral tegmental area are important in mediating the effects of opiates. This is thought to be initiated by the μ receptors present on GABA interneurones in the ventral tegmental area which inhibit GABA release, which, in turn, leads to overcoming the GABA inhibition of dopaminergic neurones and, hence, to the excitation of dopamine cells. μ Receptors are also present in the nucleus accumbens. Further support for the essential role of the μ receptor in addiction comes from recent μ receptor gene knockout experiments where the loss of morphine-induced analgesia occurs following chronic morphine treatment, as does dependence and drug withdrawal signs.

Adaptive responses are seen in opiate receptor coupling to effector systems which potentially underlie the development of tolerance

Opiate receptors belong to the G-protein-linked family of receptors containing the seven transmembrane domains shown by other members of this family such as muscarinic and adrenoceptor receptors. G-protein-linked receptors are coupled by a G-protein to an effector such as adenylate cyclase. In the case of opioid receptors, three common effects are seen:

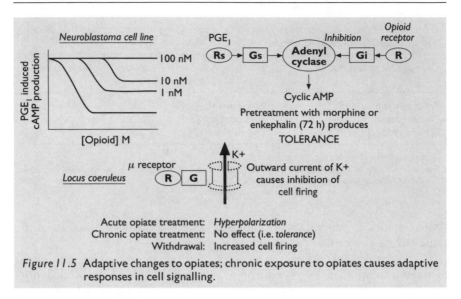

Figure 11.5 Adaptive changes to opiates; chronic exposure to opiates causes adaptive responses in cell signalling.

inhibition of adenylate cyclase, potassium channel opening and inhibition of calcium channel opening.

The inhibitory effect on adenylate cyclase represents a good model of the effect of chronic exposure to opiates. Using a neuroblastoma cell line, it has been shown that the opioid agonist inhibition of PGE_1-stimulated cyclic AMP generation is attenuated by prior exposure to opioid agonists (e.g. 72 hour pretreatment) (Figure 11.5). This represents a process of tolerance in which greater concentrations of agonists are required to produce the same effect. This effect is not due to a change in receptor number or affinity but appears to be brought about by an uncoupling between receptors and adenylate cyclase.

A similar adaptive response can be seen *in vivo* in the locus coeruleus where electrophysiological recording has been carried out in experimental animals chronically treated with opiates. In the locus coeruleus, acute administration of morphine causes inhibition of cell firing. This occurs because cells bearing the receptor are coupled by G-proteins to a potassium channel which opens, allowing potassium to leave the cell leading to hyperpolarization. After chronic exposure to opiates, this effect is diminished, whereas opiate withdrawal leads to increased cell firing (Figure 11.5). Once again, a reduced coupling is brought about by opiate exposure. ADP ribosylation of G_i is a potential mechanism whereby uncoupling could be produced.

Cannabis (marijuana, pot)

Cannabis is obtained from the flowering tops of the hemp plant, *Cannabis sativa.* Several different preparations are used, including the dried leaves and flowering shoots, the dried resinous exudate (hashish, charas) and the resinous leaves (ganja). After an oral dose (~20 mg), a state of relaxation, euphoria, altered perception of self, time and space is obtained with impaired motor co-ordination, impaired cognitive ability and spontaneous laughter, but, at high doses, hallucinations, delusions, depression and anxiety result.

THC (Δ-9-tetrahydrocannabinol) and the molecular targets of cannabinoids

Despite the fact that the analgesic and mood-altering effects of THC have been known for some time, the mode of action has only recently become apparent. The structure of THC was not determined until 1964, but, even then, the lipophilic nature of the drug hampered studies on its mechanism of action. However, with the elucidation of the structure of G-protein-linked receptors, which include muscarinic, adrenoceptor and opioid receptors (discussed above), came the discovery of one particular 'orphan' receptor which did not bind any known neurotransmitter or neuropeptide but did bind THC with high affinity and stereospecificity. The psychoactive form of THC, $(-)\Delta^9$THC, binds with a hundred times greater affinity than its stereoisomer $(+)\Delta^9$THC (Matsuda et al. 1990). The expression of the gene encoding this receptor was consistent with the distribution of THC binding, the highest concentrations being found in limbic areas such as the amygdala and hippocampus and also in the ventromedial hypothalamus and cerebral cortex. This cannabinoid receptor contains the typical features of G-protein-linked receptors possessing seven transmembrane domains, containing an N-terminal extracellular domain and an intracellular domain which interacts with a G-protein, which, in turn, transduces the effects of ligand binding to an effector system. In the case of the cannabinoid receptor, receptor activation leads to the inhibition of adenylate cyclase. The distribution of the cannabinoid receptor shows a strong association with dopaminergic pathways which are likely to be involved in mediating the euphoriant properties of THC. Lesions of dopaminergic pathways and dopamine antagonists cause an upregulation of cannabinoid receptor mRNA in striatal GABA cells which are influenced by the major dopaminergic innervation in this region.

The discovery of the cannabinoid receptor, of course, stimulated the investigation of the endogenous ligand in the same way that the opiate receptor signalled the idea of endogenous 'opiates'. Screening porcine brain extracts for their ability to displace THC in binding assays has yielded an potential candidate, an arachidonic acid derivative – arachidonyl ethanolamide (Anandamide) – which has similar behavioural properties (Devane et al. 1992). Further developments are awaited with interest.

There are a number of important potential medicinal uses for THC: in treatment of pain, inflammation, lowering intraocular pressure in glaucoma, relief of nausea in cancer chemotherapy, relief of muscle spasm in multiple sclerosis, immunosuppression and epilepsy, but the psychoactive properties have prevented use. However, the discovery of distinct peripheral receptors has opened up the possibility that some of these conditions may be treated with drugs specifically targeted at this receptor.

PCP (phencyclidine)

Phencyclidine [1-(phenylcyclohexyl)piperidine], commonly referred to as angel dust (also known as crystal, elephant, peace pill, surfer, scuffle and super weed), is a commonly used drug in the US but much less so in the UK. Originally, PCP had been developed as a clinical anaesthetic in the 1950s, the notable merits of this drug being the fact that anaesthesia was not accompanied by respiratory depression. However, the drug was withdrawn from use in humans due to the undesirable side effects, encountered when emerging from anaesthesia such as confusional states, vivid dreams, hallucinations and ataxia. The drug was replaced by a structurally similar compound, ketamine, which was shorter acting with fewer side effects, but both PCP and ketamine subsequently became drugs of abuse. In the 1970s, the incidence of schizophrenia increased several fold in the US until it was realized that the additional cases were due to PCP consumption. The schizophrenia-like symptoms characteristic of PCP abusers were depersonalization, sense of unreality and disturbance of thought processes. Understanding the mode of action of PCP is therefore relevant not only to an analysis of addiction but also, potentially, some components of schizophrenic behaviour.

The site of action of PCP and ketamine only emerged with the elucidation of the N-methyl-D-aspartate (NMDA) subtype of glutamate receptor. This is an ion channel-linked receptor which gates both Na^+ and Ca^{2+} when activated. This receptor is now known to play an important role in the process of long-term potentiation whereby, after an initial stimulus, subsequent challenges cause an enhanced synaptic response which is maintained for several hours. This potential model of memory has received considerable support both at the level of synaptic characterization and also in behavioural models. The precise organization of the NMDA receptor is not known but it is thought to be a multimeric protein whose subunits form a channel through the plasma membrane for ion transfer. The drugs – PCP and ketamine – bind to components of the channel and in this way interfere with glutamate transmission.

Good general reviews

Altman, J., Everitt, B. J., Glautier, S., Markou, A., Nutt, D. and Oretti, R. (1996) 'The biological, social and clinical bases of drug addiction: Commentary and debate', *Psychopharmacology*, **125**, 285–345.

Snyder, S. H. (1986) *Drugs and the Brain*. New York: Scientific American Books.

References

Devane, W. A., Hanus, L., Brever, A., Pertwee, R. G., Stevenson, L. A, Griffin, G., Gibson, D., Mandelbaum, A., Etinger, A. and Mechoulam, R. (1992) 'Isolation and structure of a brain constituent that binds to the cannabinoid receptor', *Science*, **258**, 1946–9.

Gawin, F. H. (1991) 'Cocaine addiction: Psychology and neurophysiology', *Science*, **251**, 1580–5.

Hoffman, B. J., Mezey, E. and Brownstein, M. J. (1991) 'Cloning of a serotonin transporter affected by antidepressants', *Science*, **254**, 579–80.

Kilty, J. E., Lorang, D. and Amara, S. G. (1991) 'Cloning and expression of a cocaine-sensitive rat dopamine transporter', *Science*, **254**, 578–9.

Kitayama, S., Shimada, S., Xu, H., Markham, L., Donovan, D. M. and Uhl, G. R. (1992) 'Dopamine transporter site-directed mutations differentially alter substrate transport and cocaine binding', *Proc. Natl. Acad. Sci. USA*, **89**, 7782–5.

Koob, G. F. and Bloom, F. E. (1988) 'Cellular and molecular mechanisms of drug dependence', *Science*, **242**, 715–23.

Kuhar, M. J., Ritz, M. C. and Boja, J. W. (1991) 'Dopamine hypothesis of the reinforcing properties of cocaine', *Trends Neurosci.*, **14**, 299–302.

Matsuda, L. A., Lalout, S. J., Brownstein, B. J., Young, A. C. and Bonner, T. I. (1990) 'Structure of a cannabinoid receptor and functional expression of the cloned DNA', *Nature*, **356**, 561–4.

Pacholczyk, T., Blakely, R. D. and Amara, S. G. (1991) 'Expression cloning of a cocaine and antidepressant-sensitive human noradrenaline transporter', *Nature*, **350**, 350–4.

Reisine, T. and Bell, G. I. (1993) 'Molecular biology of opioid receptors', *Trends Pharmacol. Sci.*, **16**, 506–10.

Rodgers, R. J. and Cooper, S. J. (1988) *Endorphins, Opiates and Behavioural Processes*. Chichester: Wiley.

Strang, J., Johns, A. and Caan, W. (1993) 'Cocaine in the UK – 1991. *Brit. J. Psychiat.*, **162**, 1–13.

Uhl, G. R. (1992) 'Neurotransmitter transporters (plus): A promising new gene family', *Trends Neurosci.*, **15**, 265–8.

Watson, S. and Girdlestone, D. (1996) *Trends Pharmacol. Sci.*, **19**(Suppl.), 51–2.

Woolverton, W. L. and Johnson, K. M. (1992) 'Neurobiology of cocaine abuse', *Trends Pharmacol. Sci.*, **13**, 193–7.

Chapter 12

Ecstasy

Marcus Rattray

Key points

- *Ecstasy is a synthetic drug, MDMA, which is an amphetamine derivative.*
- *MDMA causes psychoactive (mood altering) and psychomotor (stimulant) effects.*
- *The effects of MDMA are through release of a neurotransmitter, 5-hydroxytryptamine (5-HT), in the brain.*
- *MDMA can cause major damage to the liver and muscles, renal failure and death, chiefly through elevating body temperature.*
- *Drinking water does not prevent the organ damage caused by MDMA, and excessive water consumption can cause death through cerebral oedema.*
- *The risk of MDMA-induced death is relatively low but not predictable.*
- *MDMA may cause long-term brain damage by damaging the 5-HT system.*
- *The long-term consequences of MDMA use are unknown, but there may be increased incidence of psychiatric illness.*

Throughout the 1990s and into the twenty-first century, the drug Ecstasy has gained enormous attention. First, the drug has become extremely popular and is associated with mainstream youth culture, particularly with music and dance culture. Second, the drug has been associated with a number of fatalities, many prominently reported in the media.

Ecstasy is the popular name for a single chemical: 3,4-methylenedioxy-methamphetamine (MDMA, also known as 'Adam' or 'E', Figure 12.1). MDMA should not be confused with 'liquid Ecstasy' or 'herbal Ecstasy' as neither of these substances contain MDMA. Liquid Ecstasy is also known as GHB (γ-hydroxybutyrate) or 'GBH' (i.e. grievous bodily harm) and is a potentially lethal sedative with poorly understood effects. Herbal Ecstasy contains a naturally occurring substance called pseudephedrine

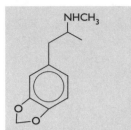

Figure 12.1 Chemical structure of 3,4-methylenedioxymethamphetamine (MDMA, Ecstasy).

which has a different structure to MDMA and effects which are more similar to the psychostimulant 'speed' (amphetamine) than MDMA. MDMA is one of a number of closely related substances known as substituted amphetamines that have broadly similar effects. These drugs include methylenedioxyamphetamine (MDA) and methylenedioxyethylamphetamine (MDEA, 'Eve') or paramethoxyamphetamine (PMA). The substituted amphetamines are entirely synthetic and were first manufactured at the turn of the twentieth century. The diversity of substituted amphetamines which can, potentially, be synthesized poses a major problem for their legal control; for example, in 2000, The Netherlands is the principal source of Ecstasy for the rest of Europe, but difficulties faced by the Dutch justice ministry in defining the chemical composition of every possible variant of the drug have hampered attempts to prevent its manufacture and interdict exports.

The 'average' Ecstasy tablet contains about 100 mg of drug, and is swallowed. Since, in most Western countries, substituted amphetamines are illegal, the manufacture of MDMA is clandestine and not subject to the rigorous quality control of legal drugs. The purity and composition of Ecstasy tablets taken by the user is usually unknown. Much of the Ecstasy found in Britain, at the time of writing, is MDA or MDEA, which are more easily synthesized than MDMA. Some Ecstasy tablets may not contain any substituted amphetamines at all, they may contain methamphetamine, LSD, ketamine or inert substances. Sometimes, tablets may contain toxic substances, although MDMA and the other amphetamines are poisonous in their own right.

About 30 minutes after ingestion of about 100 mg MDMA, the user experiences feelings of well-being, a dissociated mind state, arousal and altered perceptions akin to hallucinations. The user feels increased affinity with other people, including strangers and communication is increased. The user also displays increased motor activity similar to those found after amphetamines. Repetitive movements are common, particularly teeth gnashing and facial movements (gurning). It is likely that this altered motor co-ordination facilitates dancing, particularly to a fast, repetitive beat. The drug suppresses sleep and hunger. The initial stimulant effects of the drug typically last 4–6 hours but prolonged effects including insomnia and fatigue can occur. The subjective effects of Ecstasy can vary over time and from individual to individual – the potential range of effects has been reviewed elsewhere (Cohen 1995; Hermle *et al.* 1993). Mood depression, some days after MDMA consumption, is common (Curran and Travill 1997). These effects are likely to be due to the need for neurones to replenish their levels of a neurotransmitter called 5-HT (serotonin).

Biological actions of MDMA

Whilst MDMA and the substituted amphetamines have chemical structures that are similar to amphetamine, the psychoactive effects are markedly different. Similarities are in the psychomotor effects (increased motor activity) and inhibition of sleep. MDMA and related compounds engender a feeling of well-being that amphetamines do not. These differences relate to the specificity of these drugs to particular biochemical changes. Whilst amphetamines have a direct effect on the brain's dopamine neurotransmitter system, MDMA has its primary effect on a different neurotransmitter 5-HT (serotonin) and increases dopamine levels in the brain as a secondary effect.

MDMA and the other substituted amphetamines act specifically on the terminals of neurones (nerve cells) which make 5-HT. This is achieved because the drug binds to a protein on the membrane of the 5-HT nerve terminal called the serotonin transporter (SERT) (Rattray 1991; Rudnick and Wall 1992). The serotonin transporter is only found on 5-HT neurones and functions to recycle 5-HT, sucking the neurotransmitter back into the nerve terminal once it has been released (Figure 12.2). As well as being the site of action for MDMA, the serotonin transporter is also the site of action of a number of other drugs which are unlike MDMA, including many antidepressant drugs (including fluoxetine, Prozac) and a slimming drug (dexfenfluramine, Redux). This highlights the importance of the 5-HT neurotransmitter system in the control of mood and eating behaviour.

Once MDMA is bound to the serotonin transporter, it enters the 5-HT nerve terminal and displaces 5-HT that is stored there, causing a massive release of 5-HT. In addition, the re-uptake of released 5-HT by SERT is blocked. Because of this mode of action, MDMA is termed an 'indirect agonist' of 5-HT. The result of 5-HT release and 5-HT re-uptake block is an increased level of 5-HT in the synaptic cleft (the space between 5-HT nerve endings and other nerve cells). Studies in animals have shown that there is pronounced loss of 5-HT 30 min after MDMA, corresponding to the 5-HT release followed by subsequent rapid breakdown of the 5-HT by enzymes which are close to the site of 5-HT release. A single large dose of Ecstasy in animals can induce the release of about 20–30 per cent of the

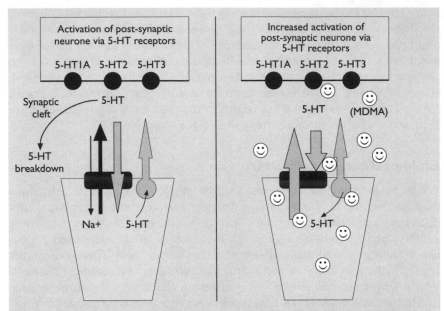

Figure 12.2 Cellular mechanism of action of MDMA. Left-hand panel represents a nerve terminal of a 5-HT neurone which is releasing 5-HT from synaptic vesicles (○) into the synaptic cleft. 5-HT binds to specific receptors (5-HT$_{1A}$, 5-HT$_2$ and 5-HT$_3$ subtypes) that are present on the membrane of neighbouring neurones (post-synaptic neurones). Through this inter-action with receptors, the post-synaptic neurones tend to become acti-vated. The 5-HT in the synaptic cleft may diffuse away and become broken down by enzymes (monoamine oxidase A), but most is taken up by the 5-HT nerve terminals via the serotonin transporter (■), in a sodium-ion dependent manner. The 5-HT is then packaged into vesicles where it can be released again. Some 5-HT is released through the sero-tonin transporter protein. The right-hand panel shows the 5-HT nerve terminal after administration of MDMA (☺). Uptake of 5-HT by the serotonin transporter is blocked. MDMA enters 5-HT nerve terminals via the serotonin transporter and displaces stored 5-HT from vesicles. Massive amounts of 5-HT are released through the serotonin transpor-ter. The result of this is greatly increased levels of 5-HT in the synaptic cleft, leading to prolonged activation of post-synaptic neurones. High doses of MDMA can remove almost all 5-HT that is stored in nerve terminals.

brain's 5-HT. The lost 5-HT can only be replaced by the increased synthesis of this neurotransmitter in 5-HT neurones. Animal studies suggest that, after a single dose of MDMA, the brain is able to replenish its store of 5-HT within 12 hours or so.

This increase in synaptic 5-HT levels is thought to account for the psy-choactive effects of Ecstasy. Although direct comparative studies between behaviour and neurotransmitter release are not yet possible in humans, the animal studies clearly support this. 5-HT which is released from nerve terminals binds to 5-HT receptor proteins that are found on numerous neurones, exciting or inhibiting them depending on the subtype of receptor. In addition, MDMA itself may bind to 5-HT receptors, particularly those of the 5-HT$_2$ type. Neurones that are activated or inhibited by MDMA are interconnected in such a way to produce the feelings of euphoria and well-being experienced by MDMA users, in a way that is not yet understood. The 5-HT released from 5-HT nerve terminals is thought to act via 5-HT receptors on dopamine neurones to produce increased release of the neurotransmitter dopamine. This is thought to underlie some of the psychostimulant (motor) effects experienced by MDMA users.

MDMA can kill

In the UK, newspaper reports suggest that at least 100,000 doses of MDMA are consumed each week, and there are several million people who have taken the drug. MDMA is popular throughout the developed world. Although the vast majority of MDMA users appear to suffer no adverse effects, there have been at least 100 deaths.

As well as the psychoactive and psychomotor effects of MDMA that are noted above, MDMA consumption causes a marked rise in body temperature, which is thought to be caused by 5-HT release in the hypothalamus, a region of the brain which controls body temperature. When combined with vigorous exercise (such as dancing), the body temperature that is achieved is similar to that which can result from a high fever and can cause profound damage to tissues (Screaton et al. 1992). The ecology of Ecstasy use is important, because Ward and Fitch (1998) have shown that use is most likely to occur at 'dance events' and, of all the chemicals which are used *while dancing*, Ecstasy is the most popular, chosen by 68 per cent of drug users in this setting. Hyperpyrexia, if left unchecked, can impair cellular homeostasis and metabolism. The Ecstasy user will collapse and become unconscious. Respiration is lowered, thus reducing oxygenation, cardiac output, lowering blood pressure and stimulating vasoconstriction. Elevated body temperature can result in a syndrome known as disseminated intravascular coagulation (DIVC), whereby the clotting factors in blood are used up; this can result in cerebral haemorrhage (although it has been suggested that the cases of cerebral haemorrhage noted in the medical literature result from amphetamine ingestion rather than MDMA ingestion: Galloway et al. 1995). Temperature elevation is considered to be the major cause of MDMA-induced fatality (Henry et al. 1992), and MDMA users who lose consciousness are usually treated by rapid cooling in ice-baths in order to minimize the damaging effects of prolonged hyperpyrexia (Matthai et al. 1996).

A consequence of elevated temperature, with increased sweating, is a loss of body fluids. The dehydration which ensues can contribute to the toxic effects of MDMA as described above. Many Ecstasy users therefore take care to drink water to replace these lost body fluids. However, excessive water consumption can itself cause damage or death through a process known as 'water intoxication'. MDMA causes release of hormones that control the excretion of sodium ions in the kidney, with the net effect to increase water retention and to lower blood sodium ion concentration (hyponatraemia) (Holden and Jackson 1996; Satchell and Connaughton 1994). Increased intake of water can further reduce the sodium ion concentration in the blood, alter blood osmolality and lead to impaired kidney function, cerebral oedema and death. Isotonic drinks (sports drinks) are certainly better than water to maintain blood ionic homeostasis, but are not a solution to hyponatraemia.

A syndrome known as rhabdomyolysis can also result from MDMA consumption (Cunningham 1997; Screaton *et al.* 1992), which may also be related to the elevation of body temperature. Rhabdomyolysis results from the increased breakdown of proteins in the body, particularly in muscle, resulting in loss of muscle mass. This can result in acute renal failure. Recent evidence has shown elevated levels of a muscle-enzyme creatine kinase in the blood of Ecstasy users, indicating that increased muscle protein turnover (Murthy *et al.* 1997). Muscle damage may be a general feature of MDMA use, even though this may not result in obvious illness. Muscle breakdown may subsequently compromise lung and kidney function (Lehmann *et al.* 1995).

In addition to the effects noted above, MDMA can also produce cardiac arrhythmias, and may therefore be particularly dangerous to those with an undiagnosed heart condition. Hepatitis (Fidler *et al.* 1996) and liver fibrosis (Milroy *et al.* 1996) have been linked to MDMA use, although how the drug causes liver disease is not known.

Thus, in addition to its mood altering and stimulant effects, MDMA can produce a wide range of damaging effects that can lead to death. Current estimates suggest about 100 deaths in the UK over the past 10 years from, perhaps, 2–3 million Ecstasy users. There are many more cases where MDMA ingestion does not prove fatal, but may leave the user with long-term health problems. It is difficult to get a true indication of the absolute or relative risk of taking MDMA compared with other drugs, but it should be noted that Ecstasy accounts for fewer deaths than alcohol, tobacco smoking and solvent abuse.

In some cases, deaths caused by Ecstasy consumption may be due to other substances in the tablets (e.g. methamphetamine), but the bulk of evidence suggests that pure MDMA can cause death. There is no indication that the dose taken or prior use of MDMA is a predictor of toxicity for an individual. Some of the people who have died were first-time users, others were thought to be regular users. Amongst the predictive factors for MDMA toxicity are the presence of cardiac abnormalities (Milroy *et al*. 1996) and whether a person has a particular type of an enzyme, cytochrome P450 (Tucker *et al*. 1994). Cytochrome P450 is an enzyme found in the liver responsible for the breakdown of foreign substances, including substituted amphetamines, and evidence has suggested that people who inherit one form of this enzyme (CYP2D6) are less able to metabolize the drug and are more likely to suffer the fatal consequences of Ecstasy use.

Effects of long-term use of MDMA

Many drugs such as nicotine (in tobacco), opioids (e.g. heroin) and cocaine are addictive, causing the user to maintain or increase their consumption of the drug over time. There is no evidence that MDMA is addictive in this way, although there are reports that habitual users of MDMA will take larger doses (e.g. seven tablets: NT Practice, 1998) in order to achieve an 'Ecstasy high'. Although there is no evidence for dependence, MDMA is defined as having a high abuse potential, since studies show that animals can be trained to self-administer the drug (Gold and Koob 1989). This 'positive reinforcing' activity is most likely related to the ability of the 5-HT released in the brain by MDMA to provoke dopamine release.

A large number of animal studies have shown that cumulative administration of two to eight high doses of MDMA over a fairly short period can cause profound alterations in the brain. Brain 5-HT levels can drop to 10–15 per cent (or less) of their normal levels, and can take up to 2 years to recover to normal levels (reviewed in Green *et al*. 1995; Rattray 1991; Steele *et al*. 1994). When the 5-HT neurones are examined under the microscope, clear evidence of damage is seen: the axons of 5-HT neurones (nerve fibres) are broken and swollen. Enzymes that are known to be present in healthy 5-HT nerve terminals are also lost. These studies show that MDMA can be neurotoxic, and this toxicity is selective to 5-HT neurones. 5-HT neurones do not actually die, but their extensive branched network of axons is severely pruned. Studies by George Ricaurte and his colleagues suggest that the axons gradually grow back, but are not able to reach some of the brain regions which are normally rich in 5-HT axons (Fischer *et al*. 1995).

It is not yet clear to what extent these neurotoxic effects occur in humans. At the time Ecstasy became popular in the 1980s, it was not yet possible to survey damage to the 5-HT system in the living human brain, although studies in monkeys showed evidence of 5-HT neuronal damage in the living brain (Scheffel *et al*. 1998). Indirect measurements showed that chronic MDMA users have lower levels of 5-HT metabolites in their cerebrospinal fluid, suggestive of a compromised 5-HT system. These studies were confirmed somewhat by endocrinological assessment (the prolactin response), an indirect assay of the integrity of the central 5-HT system

(McCann *et al.* 1994). Now positron emission tomography of the human brain has demonstrated the selective disruption of the 5-HT system in Ecstasy users (McCann *et al.* 1998).

These studies are, of course, complicated by the fact that many chronic MDMA users are frequent users of other drugs, and that the amounts of drug taken by each user is unknown and will vary from person to person.

Since the 5-HT system is important for the control of mood and emotion, it is possible that individuals with a compromised 5-HT system may experience psychiatric illness. Depressive illness, for example, is thought to relate to low levels of 5-HT in the brain, whilst anxiety is thought to relate to elevated 5-HT levels. In the short term, MDMA may precipitate panic attacks (Pallanti and Mazzi 1992) and depression (Curran and Travill 1997), presumably due to release of 5-HT and temporary depletion of 5-HT, respectively. There are numerous case reports that depression and other forms of psychiatric illness, particularly psychosis, are associated with MDMA use (Cohen and Cocores 1997; McGuire *et al.* 1994). Sleep abnormalities in long-term Ecstasy users have also been reported (Allen *et al.* 1993). However, it is clear that for the vast majority of MDMA users there are, so far, few obvious effects of the drug. One cause of concern, however, is that if MDMA damages the brain's 5-HT system, there may be an increased risk of psychiatric illness later in life.

References

Allen, R. P., McCann, U. D. and Ricaurte, G. A. (1993) 'Persistent effects of (+/−)3,4-methylenedioxymethamphetamine (MDMA, 'ecstasy') on human sleep', *Sleep*, **16**, 560–4.

Cohen, R. S. (1995) 'Subjective reports on the effects of the MDMA ('ecstasy') experience in humans', *Prog. Neuropsychopharmacol. Biol. Psychiat.*, **19**, 1137–45.

Cohen, R. S. and Cocores, J. (1997) 'Neuropsychiatric manifestations following the use of 3,4-methylenedioxymethamphetamine (MDMA: 'Ecstasy')', *Prog. Neuropsychopharmacol. Biol. Psychiat.*, **21**, 727–34.

Cunningham, M. (1997) 'Ecstasy-induced rhabdomyolysis and its role in the development of acute renal failure', *Intensive Crit. Care Nurs.*, **13**, 216–23.

Curran, H. V. and Travill, R. A. (1997) 'Mood and cognitive effects of +/−3,4-methylenedioxymethamphetamine (MDMA, 'ecstasy'): Week-end 'high' followed by mid-week low', *Addiction*, **92**, 821–31.

Fidler, H., Dhillon, A., Gertner, D. and Burroughs, A. (1996) 'Chronic ecstasy (3,4-methylenedioxymethamphetamine) abuse: A recurrent and unpredictable cause of severe acute hepatitis', *J. Hepatol.*, **25**, 563–6.

Fischer, C., Hatzidimitriou, G., Wlos, J., Katz, J. & Ricaurte, G. (1995) 'Reorganization of ascending 5-HT axon projections in animals previously exposed to the recreational drug (+/−)3,4-methylenedioxymethamphetamine (MDMA, 'ecstasy')', *J. Neurosci.*, **15**, 5476–85.

Galloway, G., Shulgin, A. T., Kornfeld, H. & Frederick, S. L. (1995) 'Amphetamine, not MDMA, is associated with intracranial hemorrhage', *J. Accid. Emerg. Med.*, **12**, 231–2.

Gold, L. H. and Koob, G. F. (1989) 'MDMA produces stimulant-like conditioned locomotor activity', *Psychopharmacology (Berl.)*, **99**, 352–6.

Green, A. R., Cross, A. J. and Goodwin, G. M. (1995) 'Review of the pharmacology and clinical pharmacology of 3,4-methylenedioxymethamphetamine (MDMA or 'Ecstasy')', *Psychopharmacology (Berl.)*, **119**, 247–60.

Henry, J. A., Jeffreys, K. J. and Dawling, S. (1992) 'Toxicity and deaths from 3,4-methylenedioxymethamphetamine', *Lancet*, **340**, 384–7.

Hermle, L., Spitzer, M., Borchardt, D., Kovar, K. A. and Gouzoulis, E. (1993) 'Psychological effects of MDE in normal subjects. Are entactogens a new class of psychoactive agents?', *Neuropsychopharmacology*, **8**, 171–6.

Holden, R. and Jackson, M. A. (1996) 'Near-fatal hyponatraemic coma due to vasopressin over-secretion after 'ecstasy' (3,4-MDMA)', *Lancet*, **347**, 1052.

Lehmann, E. D., Thom, C. H. and Croft, D. N. (1995) 'Delayed severe rhabdo-myolysis after taking 'ecstasy'', *Postgrad. Med. J.*, **71**, 186–7.

McCann, U. D., Ridenour, A., Shaham, Y. and Ricaurte, G. A. (1994) 'Serotonin neurotoxicity after (+/−)3,4-methylenedioxymethamphetamine (MDMA; 'Ecstasy'): A controlled study in humans', *Neuropsychopharmacology*, **10**, 129–38.

McCann, U. D., Szabo, Z., Scheffel, U., Dannals, R. F. and Ricaurte, G. A. (1998) 'Positron emission tomographic evidence of toxic effect of MDMA ('ecstasy') on brain serotonin neurons in human beings', *Lancet*, **352**, 1433–7.

McGuire, P. K., Cope, H. and Fahy, T. A. (1994) 'Diversity of psychopathology associated with use of 3,4-methylenedioxymethamphetamine', *Brit. J. Psychiat.*, **165**, 391–5.

Mallick, A. and Bodenham, A. R. (1997) 'MDMA induced hyperthermia: A survivor with an initial body temperature of 42.9 degrees C', *J. Accid. Emerg. Med.*, **14**, 336–8.

Matthai, S. M., Davidson, D. C., Sills, J. A. and Alexandrou, D. (1996) 'Cerebral oedema after ingestion of MDMA ('ecstasy') and unrestricted intake of water', *Brit. Med. J.*, **312**, 1359.

Milroy, C. M., Clark, J. C. and Forrest, A. R. (1996) 'Pathology of deaths associated with 'ecstasy' and 'eve' misuse', *J. Clin. Pathol.*, **49**, 149–53.

Murthy, B. V., Wilkes, R. G. and Roberts, N. B. (1997) 'Creatine kinase isoform changes following Ecstasy overdose', *Anaesth. Intensive Care*, **25**, 156–9.

NT Practice (1998) 'Beliefs the key to getting anti-ecstasy message across', *Nursing Times*, **94**(48), 46.

Pallanti, S. and Mazzi, D. (1992) 'MDMA (Ecstasy) precipitation of panic disorder', *Biol. Psychiat.*, **32**, 91–5.

Rattray, M. (1991) 'Ecstasy: Towards an understanding of the biochemical basis of the actions of MDMA', *Essays Biochem.*, **26**, 77–87.

Rudnick, G. & Wall, S. C. (1992) 'The molecular mechanism of 'ecstasy' [3,4-methylenedioxy-methamphetamine (MDMA)]: Serotonin transporters are targets for MDMA-induced serotonin release', *Proc. Natl Acad. Sci. USA*, **89**, 1817–21.

Satchell, S. C. and Connaughton, M. (1994) 'Inappropriate antidiuretic hormone secretion and extreme rises in serum creatinine kinase following MDMA ingestion', *Brit. J. Hosp. Med.*, **51**, 495.

Scheffel, U., Szabo, Z., Mathews, W. B., Finley, P. A., Dannals, R. F., Ravert, H. T., Szabo, K., Yuan, J. and Ricaurte, G. A. (1998) 'In vivo detection of short- and long-term MDMA neurotoxicity – a positron, emission tomography study in the living baboon brain', *Synapse*, **29**, 183–92.

Screaton, G. R., Singer, M., Cairns, H. S., Thrasher, A., Sarner, M. and Cohen, S. L. (1992) 'Hyperpyrexia and rhabdomyolysis after MDMA ('ecstasy') abuse', *Lancet*, **339**, 677–8.

Steele, T. D., McCann, U. D. and Ricaurte, G. A. (1994) '3,4-Methylenedioxy-methamphetamine (MDMA, 'Ecstasy'): Pharmacology and toxicology in animals and humans', *Addiction*, **89**, 539–51.

Tucker, G. T., Lennard, M. S., Ellis, S. W., Woods, H. F., Cho, A. K., Lin, L. Y., Hiratsuka, A., Schmitz, D. A. and Chu, T. Y. (1994) 'The demethylenation of methylenedioxymethamphetamine ('ecstasy') by debrisoquine hydroxylase (CYP2D6)', *Biochem. Pharmacol.*, **47**, 1151–6.

Ward, J. and Fitch, C. (1998) 'Dance culture and drug use', in G. V. Stimson, C. Fitch and A. Judd (eds) *Drug Use in London*, pp. 108–4. London: Leighton Print.

Chapter 13

Adolescent drug use and health

Problems other than dependence

Woody Caan

Key points

- *A wide variety of psychoactive substances are available to young people around the world, millions of whom are taking drugs occasionally or periodically in ways which do not reveal a dependence on drug use.*
- *It is a matter of serious concern to the caring professions when they are confronted with the harm resulting from drug use in a young person or when they discover hazardous behaviour which carries a great risk of future harm.*
- *Children and adolescents may rely on more experienced adults, especially those with a professional 'label', to act as models of behaviour and sources of advice.*
- *It may be helpful to anticipate the choices a young person might make about drug use or non-use within a longer range view of their health and emotional development.*
- *Problems arise in different ways: toxic actions of the drug (e.g. sudden death while inhaling butane lighter fluid); hazardous intake of the drug (e.g. injecting hepatitis viruses along with heroin); dangerous behaviour under the influence (e.g. driving after smoking cannabis); after-effects of use (e.g. the emotional 'crash' after a binge of taking cocaine); cumulative toxicity (e.g. panic attacks and depression associated with Ecstasy); secondary casualties (e.g. HIV infection of a foetus after its mother shares an unsterile injection of buprenorphine).*
- *It is important to remember that most of the people at risk of these problems are NOT burdened by an addiction to their drug(s) and are likely to respond to educational and social initiatives in similar ways to other adolescents.*
- *An appreciation of the context of using drugs – in that individual's life story, their personal level of understanding about risk and the setting in which drugs are available (e.g. a party or a prison) – is a helpful start to preventing any drug-related problem.*

Most people who take illicit ('street') drugs are not 'addicts' displaying dependence on these substances. Patterns of illicit drug use principally begin with 10- to 19-year-olds, and, in their 1997 study of 1,000 British parents, NCH Action for Children identified drug and alcohol abuse as the most common fear for their children, with 89 per cent of parents expressing concern. Teenage drug use is a highly emotive subject in most societies, for all sections of society, including children and professionals, so I will begin with an illustration of hazardous drug use from a different generation, in a 'protective' environment and with licit substances.

A lesson from care homes for older people

Older people are most commonly introduced to sleeping pills (such as the benzodiazepines temazepam or lorazepam) by doctors after one of two sorts of life event: bereavement or an acute hospital admission. Disturbed sleep is common during difficult periods of adjustment (such as after loss of a husband) and is actually a diagnostic feature of adjustment disorder in the current ICD-10 system. Well-meaning family doctors may prescribe hypnotic drugs, although there is no evidence of therapeutic benefit, even in clinical cases of adjustment disorder. In a busy and short-staffed hospital ward, night sedation may be prescribed after only cursory medical consideration and continued for long periods after discharge, even though any effect on prolonging sleep disappears within 4 weeks of regular use. Back in their community, the old person who has been started on one sedative is quite likely to be given additional types of medication, and to drink alcohol, adding to the hazards of polypharmacy. In the population of older people, hypnotic drug use is a major risk factor for falling (Di Fabio and Seay 1997): benzodiazepine use raises the danger of multiple falls to more than seven times the level of risk seen in non-users (Lord *et al.* 1995). Such falls are especially frequent in a particular environment which includes many residents with life histories of bereavement and past hospital admissions: residential homes for the elderly (DTI 1990). Falls might result directly from the effects of nightly drug *intoxication* on a resident's motor co-ordination and judgement of danger, or from an altered perception and attention during the drug *hangover* in the mornings, or consequent to *cumulative* effects of the drug producing forgetfulness and disorientation. Some individuals may react to psychotropic drugs in idiosyncratic ways, such as sudden mood swings or aggression towards other residents. Older people may even end up falling over each other: consider the notorious television spectacle of a crowded dining table where an elderly war veteran using triazolam (US President George Bush) collapsed over the Prime Minister of Japan (a *fallout* of benzodiazepine misuse can lead to *secondary casualties*). So, once the residents of a home start falling over, their confidence drops and their activities become restricted, and an already disadvantaged group becomes further disabled. Many older people are developing osteoporosis ('brittle bones') with ageing and these bones break when they fall, especially at the hip. The scale of this danger is seldom apparent to the lay public, but among the over-65s in East Anglia, fractured neck of femur is by far the largest *preventable* cause of accidental disability, financial burden to public services and premature death. In summary, a well-meaning physician can make an elderly patient's life significantly more hazardous by the innappropriate supply of drugs, just as if they were scattering banana skins in their path, instead of benzodiazepines.

In approaching potentially dangerous substance use, the responsibility of *all* professionals is to identify the nature of the risks, and, if significant problems are foreseeable, either to help clients reduce their hazardous drug use or to encourage use in a safer way. When caring for young people below the statutory age of majority, in our contemporary society which offers myriad choices to adolescents, skill and sensitivity are particularly needed:

- in assessing individual patterns of drug taking;
- in gauging the risks in the context of that person's life; and
- in promoting health and personal development for them.

The most popular drugs (this year)

The picture of initial use and choice of drugs varies over time, but the Commission of the European Communities (1995) has identified the following *trends* across Europe:

1   an increase in the number of first time users;
2   an increase in drug-related mortality and morbidity;
3   an increase in the number of drug-related AIDS cases.

Within the European Union in 1997, the European Monitoring Centre for Drugs and Drug Addiction has identified *Britain* as having the most young adults using illicit drugs, especially cannabis, amphetamine, Ecstasy and LSD. The HEA (1997a) suggested that in 1996 the UK's most popular drugs in the age range 16–19 were cannabis and hallucinogens (like LSD and Ecstasy). Over all ages, 16 per cent of Englishmen had used cannabis in the last year and 8 per cent used hallucinogens, but these average percentages were both *doubled* for boys aged 16–19, and drug use was even higher in girls aged 16–19 (cannabis, 40 per cent and hallucinogens, 21 per cent). In particular populations, we can observe 'fashions' for a certain drug. In 2000 for example, a survey by the London magazine *Time Out* found that 45 per cent of its readers had tried cocaine powder and another 6 per cent had smoked crack cocaine (Bloomfield and Kerr 2000). Other, more systematic investigations have confirmed a pattern of increased cocaine use among 16- to 24-year-olds, localized within London (Corkery 2000). Across Britain, Miller and Plant (1996) studied over 7,000 adolescents aged 15–16, and found significant differences in the pattern of use between geographical regions within one country; for example, 16 per cent of Scottish boys used tranquillizers, more than three times the rate in English, Irish and Welsh boys, whereas the boys in Northern Ireland had a higher likelihood of using glues and solvents (28 per cent) than adolescents in any other regions. However, for both sexes and all regions, one consistent pattern emerged: *higher levels of drug use were associated with poorer school performance.*

Teenagers show a number of linked behaviours, which have been described since 1989 in a series of Health Education Authority studies involving tens of thousands of young people; for example, only 2 per cent of 13- to 15-year-olds who have never smoked cigarettes have tried any illicit drug, but 52 per cent of those who are regular smokers have also tried drugs like cannabis, LSD or Ecstasy. Across all drugs, smoking is the most popular route of administration (including smoking crack cocaine and 'chasing the dragon' with heroin), but few children smoke crack before first having lit a cigarette. If young people smoke, the age of starting is important. During a study of Oregon schoolchildren, those who had started smoking early (e.g. before 11) were more likely to develop problems related to drug use between the ages 19–24 than smokers who started later (Lewinsohn *et al.* 1999). In the USA, a comparable series of observations called the Youth Risk Behaviour Surveillance System has been tracking phenomena like cannabis use during the school years and relating a wide range of adolescent behaviours to the four events (car crashes, other unintentional injuries, homicide and suicide) which together account for 72 per cent of all deaths in this age group.

Drug 'fashions' vary significantly with time as well as by region. A joint project by the Adfam national helpline and BBC Radio beginning in 1989 saw a striking rise in 1996 of concerns related to heroin use (29 per cent of calls) and cocaine use (13 per cent of calls) under the age of 18. Among *Time Out* readers (Bloomfield and Kerr 2000), 24 per cent reported 'collapsing or passing out' under the influence of drugs. During the 1990s, adolescent risk taking *in general* has risen significantly (Calman 1994) and in parallel Roberts *et al.* (1997), at London's Institute of Child Health, found that death rates from poisoning rose sharply for opiates like heroin (from 17 teenagers between 1985–89 to 67 between 1991–95) and for stimulants like amphetamine or cocaine (from 1 teenager in 1985–89 to 16 in 1991–95).

Hazardous and harmful use

Injury and poisoning are by far the most common causes of death observed in young people between the ages of 10–19, with more fatalities among boys than girls. Between 1986 and 1996, both male and female deaths attributed to poisoning by drug misuse in England and Wales rose across all age groups (Calman 1997). Among boys aged 15–19, there was a 109 per cent increase in drug-related deaths, and the risk of such fatal poisoning rises to a peak between the ages of 25–29 years. In a country town the size of Cambridge, this national increase is equivalent to one or two *additional* drug deaths per year, every year, of young adults 'in their prime'. Recently, the UK Government launched an Action Plan to prevent such drug-related deaths (Drugs Misuse Team 2001). Considering the hazards from large numbers of young people trying various street drugs, *deaths* from 'recreational' drug use are fortunately not everyday events, but other types of *harm* are becoming more and more apparent, such as drug use undermining the mental health of many young people (Caan 1997a).

Public health experts in Cambridge are becoming increasingly aware that most adults have only very vague concepts of 'risk' (Zimmern 1997). The absolute risk of death in every year of intravenous drug use is one in fifty. This is thirty times more likely to prove fatal to a 17-year-old than taking up mountain climbing as a hobby, and 160 times more likely than a fatal road traffic accident if they were to take up driving. But 17-year-olds probably *see themselves* as invulnerable ... yet in one much-studied American community sample, who began using a variety of substances in adolescence, 90 per cent of those who had ever injected drugs intravenously reported some related physical, psychological or social harm (Dinwiddie *et al.* 1992a,b).

A taste of danger ...

One November, during a visit to a treatment centre where I was researching the way dealers market heroin and cocaine to users, the nurse in charge was bidding good evening to an informal group of drug injectors as they drank coffee together. When that nurse mentioned that his brother ran a fireworks business and he was on his way to help him give a public display, all the users became agitated, asking him not to do anything SO DANGEROUS ...

As a caring professional, I tend to see a population of drug users who *present* with definite physical, psychological or social *problems*, and I attempt to use professional judgement about the extent to which drugs cause or aggravate *their* problems. Most adolescent users never contact a drug service, and very localized factors like their friends and families dominate their *subjective* view of drug taking (Health Advisory Service 1996; Demos 1997) which may well be positive or neutral overall. In their survey of drug use in the UK's 'rave' dance culture, Release (1997) found that 97 per cent of the dancers had tried illicit drugs sometime and 87 per cent intended to use a drug at that night's dance. Although some reported negative experiences (e.g. 28 per cent of the LSD users had experienced 'feelings of paranoia'), the majority of these recreational users had positive views that their drug of choice brought them benefits like 'energy', 'confidence' or 'relaxation'. Four-fifths of the drug-using 15- to 16-year-olds in Wales took their drugs with friends (Roberts *et al.* 1995), although there was a gender difference in that the girls were more likely to take drugs 'when they were feeling down' (33 per cent did this – compare that with the female use of alcohol in Chapter 6). Occasional cocaine users interviewed in Scotland mostly described beneficial effects like 'You're outgoing, no shyness, you talk to anybody' and 'It's, let me apply myself a wee bit more . . . no distractions: concentrate a wee bit better . . . you seem to home in on things better' (Bean 1993). Some substances seem to have a surprising degree of social acceptance – my Sunday paper today has a front page headline 'Want some dope'? Try the Houses of Parliament' about the supply of cannabis there.

However, if you are a family doctor on call for home visits or a community nurse manning a telephone helpline or in an urban crisis centre as a member of a psychiatric emergency team, a major part of *your* work involving drugs will be attending to young people experiencing distressing effects of recent substance use, often accompanied by frightened relatives (Caan 1997b). Sometimes, you will not be able to determine what substance or mixture of substances has been taken, but certain types of drug are likely to come up, year after year. Hallucinogens like LSD ('Acid') or ketamine ('Special K') can often lead to scary 'Bad Trips', and a heavy run of using stimulants like amphetamine ('Speed') or cocaine ('Snow' or 'Crack') can put adolescents on what appears like a wild emotional roller coaster. Some methylated amphetamines like MDA and MDMA ('Ecstasy') combine the features of *both* hallucinogens and stimulants (usually with an added variety of unknown adulterants as well) giving highly variable effects (Caan 1997a). So, for a psychiatric liaison nurse in an inner city Accident and Emergency department:

> the most common difficulties are related to substance misuse.
>
> (Watts, 1997)

Is there a framework to help us understand drug problems in youth and to interpret the possible hazards?

Intoxication

The physics and chemistry of the drug itself can cause harm; for example, some children misuse butane lighter fluid or aerosol propellants to produce a partially anaesthetized state, but die as a consequence (Anderson 1990). When inhaled anaesthetics were first being developed in the nineteenth century, 'laughing gas parties' were sometimes held to generate a pleasantly light-headed or dreamy state, and in the twentieth century children in many countries were observed to seek a similar intoxication from sniffing petrol or the solvent toluene in glue. Glue sniffers had been observed to die from choking on their vomit after entering too deep a level of anaesthesia or, if they combined a lighter anaesthesia with childhood enthusiasms like climbing trees, a fatal accident could then ensue. With the recent fashion for aerosols and lighter refills, sudden death has become more common – it only takes one attempt to squirt the gases directly into the mouth to freeze the air passages, and heart failure may also bring an abrupt end to a session of inhalant use. Most inhalants are flammable and accidental burns are another risk: a 12-year-old savouring cigarette lighter fluid and also lighting cigarettes in the same room makes a hazardous combination.

Accidents while intoxicated are the most common problem seen with cannabis, and you do not have to be as young as 12. Howard Marks (1997), the leader of the current UK campaign Decriminalise Cannabis, reports 'it's very irritating for my wife, but I actually fall asleep with a joint and I like waking up in the morning to find that I'm sleeping on a joint'. Cannabis is frequently recorded in the charred remains of fire victims (e.g. one of my student friends came to an abrupt end trying to use a gas cooker while intoxicated) and one wonders what the outcome for Mr and Mrs Marks will be of his smoking in bed every night. As well as localized studies of cannabis as a factor in deaths by, say, fire and drowning, several large-scale studies have suggested that this drug is second only to alcohol as a factor in fatal road accidents (e.g. 10 per cent of 1997 deaths in a conference report from the UK Department of Environment and Transport), and, compared with 10 years ago, illicit drugs are now appearing in four times as many cadavers from road traffic accidents.

Overdoses are a common problem when CNS depressants which impair breathing and lower body temperature are tried in uncertain 'street' doses or when taken accidentally by small children in a drug-using household. In one Australian study based on a coroner's sample of 231 poisonings, 40 per cent had taken the opiates heroin, morphine or methadone (Coleridge et al. 1992). Fortunately, there is a good antidote to such opiates, Naloxone, and the great majority of heroin overdoses that reach emergency medical services before death will then survive. An overdose prevention initiative giving heroin users themselves a Naloxone supply at home, to help their peers who overdose, is being tried now (Phillips and Johnson 2000). In the USA, cocaine (especially when mixed with alcohol, producing the toxic metabolite cocaethylene) is the most common component of fatal poisonings. Although there is no available antidote for cocaine overdose (which may be accompanied by cardiovascular problems, stroke and fits:

Strang *et al.* 1993), most cases that reach an emergency service promptly survive with supportive care. The substances commonly seen in poisonings can vary dramatically over time; for example, Russians have been drinking vodka for centuries, but recently there has been an alarming increase in alcohol poisoning – producing about five times more deaths among 15- to 19-year-olds (Leon *et al.* 1997). The pharmaceutical industry has developed an antidote for benzodiazepine poisoning, Flumazenil, and benzodiazepines do figure in many overdoses (40 per cent in the Australian study: Coleridge *et al.* 1992). However, more than five out of six such fatal poisonings involved other drugs combined with the benzodiazepines, and we do not yet know the safety or effectiveness of Flumazenil in 'mixed' overdoses. The history of so-called 'analeptic' drugs, which were used in the 1950s and 1960s to counteract barbiturate overdose, offers a note of caution: analeptics killed many more patients than they helped and were then abandoned in favour of other, safer approaches.

- The crucial message in all drug overdoses is *prompt* access to Accident and Emergency services.

In some human and animal observations, a single dose of hallucinogens (especially phencyclidine and ketamine) can cause neurological damage (Caan 1996a). Idiosyncratic responses by individuals to a multitude of substances have been reported; for example, several hours of frightening hallucinations after eating a nutmeg pudding (McGrath 1997). In young people, recent drug intoxication may sometimes cause a spectrum of severe anxiety, hallucinosis and delusional disorders, even at 'ordinary' or 'customary' doses. It is claimed that 20 per cent of UK university students are regular cannabis users (Swain 1997) which may explain why this author is asked so often to lecture on 'cannabis psychosis'. Below, I will consider cannabis use and mental health briefly, but *within the context* of an overview of drug use and mental health.

Drug use and the risk of psychological harm

It seems amazing that a recent Editorial in the *British Journal of Psychiatry* advises that the term 'drug induced psychosis' is unproven and unsustainable (Poole and Brabbins 1996), given that there has been an abundant literature on this field over at least 90 years, but then these authors fail to reference a single study in German or French, in which languages most of the relevant literature was written. Emil Kraeplin, the father of modern psychiatry, revised his classic textbook (Kraeplin and Lange 1927) to include a chapter on *Kokainismus*, a term which included an epidemic of both cocaine-induced mental disorders and cocaine dependence around the end of the First World War. Given that Kokainismus had risen from 0 per cent of German psychiatric admissions in 1915 to 13 per cent of all admissions by 1924 (Petersen 1978), one can understand a need for revising the textbooks. I was privileged to know the late pharmacologist Dr Hermann Blaschko, who served then as a junior doctor in the world's first specialist cocaine unit, and the memory of trying to treat a gang of armed and deluded young soldiers returning from the Eastern Front remained with him vividly into his 90s. Readers with a special interest in cocaine effects may want to look up Lewin (1924), Maier (1926) and Deschamps (1932). Hermle et al. (1993) offer a simple summary of German experimental approaches to a wide range of substance-induced psychoses.

The best descriptions of drug-related mental disorders come from experimental research with healthy young people. The Edgewood studies (finishing in 1975) tested hallucinogens including LSD and phencyclidine on US servicemen for their potential in chemical warfare. While intoxicated, most of the men were effectively disabled as soldiers and a minority developed florid and fearful conditions shortly after administration of the hallucinogens (Bad Trips). Residual mental problems did persist *after* the drug had been cleared from the body in a few cases (not all subjects were followed up, so we will never know the percentage overall). The next best evidence comes from observations after the use of a drug as a therapeutic agent with non-psychotic patients. Given as an anaesthetic, ketamine can sometimes be observed to produce abnormal mental states, with a risk of reckless (self-harming) behaviour (Jansen 1993). Amphetamines given to asthmatics or narcoleptics sometimes gave rise to *de novo* mental disorders, although this may only have appeared after a number of doses (stimulants can be observed in animals to have priming or kindling effects when given repeatedly, even at low doses). The patients in Connoll's (1958) description of amphetamine psychosis at the Maudsley Hospital had generally been taking amphetamines for some time before they presented as psychotic. A generation earlier, the hallucinogen mescalin was being tested on hundreds of Maudsley patients as a therapeutic agent to improve creative expression, and this sort of clinical study (with cocaine, amphetamine, MDA or LSD) has generally shown that some patients relapse or develop extra pathology when given hallucinogens or stimulants as a 'tonic'. In 2000, the High Court in England heard the personal injury claims of eighty-seven former patients from twenty services who 'treated' them with LSD, 30 to 40 years ago.

All the above types of evidence at least allow us to ensure the dose and purity of the drug and observe the effects before, during and after use. Unfortunately, most descriptions of psychological 'harm' come from clinical case studies, where all we see is the patient's condition after use. Allen and Jekel (1991) give a clear, English-language description of an acute onset psychosis after high-dose use of crack in the Bahamas, involving terrifying hallucinations and a sense of doom – this sounds similar to Wolff (1945) remembering cocaine injectors 'brought to hospital in a state of intoxication, fear or hallucination' with 'painful delusions' and 'extreme agitation' suggesting a case of 'acute cocaine poisoning'. Allen and Jekel suggest that such patients recover quickly, often within 24 hours, if they do not take any more cocaine. Note that, unlike the reaction to a single dose of nutmeg pudding above, most incidents of 'cocaine psychosis' or of a similar disorder related to 'epidemic' methamphetamine use in Japan after World War Two have been described in people who have *frequently* taken stimulants *before* the occasion of their psychotic episode. With very long-term stimulant use an 'insidious onset psychosis' (Allen and Jekel 1991) has also been described involving bizarre foraging and collecting behaviours in a context of fretful, preoccupied worrying. With intoxicated users who had developed dependence on either amphetamine or cocaine, I have observed intense collecting and treasuring of pebbles and gravel, which may be what these authors mean by 'ritual' foraging. Where case studies of users experimenting with drugs *for the first time* have described mental consequences, the descriptions often include panic attacks; for example, The Phobics Society have described this occurring in some young people shortly after either Ecstasy or cannabis use.

So where does cannabis fit into this picture? The review by Ashton (1987) concluded that a psychosis could occur after 'acute exposure even to moderate doses of cannabis and Δ^9-tetrahydrocannabinol, especially in those not used to taking the drug'. A history of frequent cannabis use (fifty times or more) *predicted* a sixfold increased risk of developing a schizophrenic psychosis in a follow-up study of more than 45,000 Swedish soldiers (Andreasson *et al.* 1987), heavy cannabis use *preceded* psychotic symptoms in the clinical sample of Linzsen *et al.* (1994), and the study of first psychiatric admissions by Adams and Silverstone (1991) showed that recent cannabis use was a *better* predictor of having a psychotic diagnosis than other types of substance use by their clientele. A positive urine test for cannabis was strongly associated with psychotic symptoms *at the time* of admission, compared with subjects without cannabis in their urine, in our clinical study of 908 patients (Mathers *et al.* 1991). A follow-up study of corresponding psychotic subjects, with and without cannabis use, found that cannabis was associated with a *more severe* schizophrenic illness, at first assessment, than was typical of the non-using patients.

Compared with LSD, the evidence regarding cannabis is weaker. Considering the large numbers of young people smoking cannabis, any resultant mental disorders seem relatively rare. However, from the point of view of the caring professions, in *those populations likely to be in contact with psychiatric services* cannabis use might well predispose to the onset of mental illness and probably does make the symptoms of people with a pre-existing illness more severe. As genetically selected breeds of the plant *Cannabis sativa*, which yield consistently higher doses of the psychoactive tetrahydrocannabinols like

'Skunk Weed', take over the market from traditional varieties, any such *risks* from intoxication with cannabis (Smith 1998) are likely to be amplified.

What might the abnormal psychological states induced by drugs be like?

A subjective view

Sometimes, it is difficult to communicate the disturbing experience of a drug-induced bad trip. A certain boy had great difficulty describing the delayed effect of one hallucinogen to his parents, in which a pack of large grey wolves suddenly emerged from one wall of the house, then swarmed nightmarishly around as they galloped across the stairs, until finally the wolves vanished in sequence, nose first, into a far wall. All a parent saw was their son briefly petrified with fear and later refusing to climb the stairs. On a later occasion, eating hallucinatory mushrooms with a school friend seemed to produce no effect at first except an abdominal ache and some queasiness, for which (after perhaps 2 hours including a short drive) that young man eventually lay down. Suddenly, an intense vestibular sensation like a lift rocketing up to the ceiling, followed by the visual perception of the walls melting and flowing around him and a head-to-toe tactile sensation of waves bending along his body, culminated within seconds in a snowstorm of sensations in which movement, sight and touch were no longer distinct dimensions but confusingly overlapped ('synaesthesia'). Time appeared to stand still, so subjectively he had no idea how long this dreadful synaesthesia snowstorm lasted, but, subsequently, he remained much too agitated to sleep that night and periodically lost the sense of any boundaries between where his body ended and the ceiling, walls and bed began: not so much an 'out of body' experience as an 'in-house' sensation. The next morning, his school friend telephoned to say that, despite taking the same drug, he had had no psychotomimetic experience, just anxiety, much vomiting and a sleepless night. Today, that user feels so grateful that the abrupt hallucinatory episode did not begin an hour *earlier*, while he was trying to drive a car.

Objective problems

A 17-year-old male visited while experiencing the fluctuating perceptual changes induced by LSD. In the middle of a rambling conversation, he was emptying the contents of his shaving bag and picked up a razor blade, with which he began nonchalantly cutting his forearm. Rather than perceiving this experience as painful, he became fascinated with the strange appearance of the blood issuing from his cuts, which he repeated more and more deeply to view the unusual ('beautiful') colours and motions of the blood as it fell towards the floor. He expressed no alarm when his razor was removed and the bleeding was staunched, only regret that the visual display was over. Fairly soon after this, he began his medical studies.

A study of deliberate self-harm ('DSH') in young Oxford males aged 15–24 has reported an increase of 194 per cent in DSH between 1985 and 1995, often associated with their use of drugs (Hawton *et al.* 1997).

One winter night, an 18-year-old male student from Africa with no previous psychiatric history, who was inexperienced in substance use, tried a hallucinogen (reportedly LSD) in the privacy of his room. He became convinced that he was in great danger, pursued relentlessly by hostile soldiers across the bushland of his boyhood. Running out of his lodging into falling snow and bitter cold in his shirtsleeves, he felt no cold and held the view that his surroundings were in dazzling tropical sunshine, with his army of pursuers all around him. After a prolonged period of distressed yelling and running around the snow, sometimes attempting to hide behind objects (perhaps trees, but invisible to us), university staff came to his aid and he was taken to a psychiatric hospital as a place of safety, where his condition merited admission under a section of the Mental Health Act. For a couple of days, he remained very fearful and distressed, but the experienced hospital staff prevented any physical harm to the young man during this vulnerable state and within a week the hallucinations and delusions had ended.

Cawley *et al.* (1997) have found that 'drug abuse' is the most common antecedent of brief psychiatric admissions (26 per cent of all the admissions lasting between 1 and 7 days), with the average drug user 17 years younger than the average patient with schizophrenia who had a similar brief admission. Reducing drug use was the key factor in reducing psychiatric re-admissions in a study that linked the community environment to hospital use or imprisonment (Parkman 2000). Clients saw quality of life dimensions like social support and economic security as more important than their drug use, and so an effective community service to prevent mental health crisis should combine the goals of less drug use and improved social circumstances.

Hangover

The best described after-effect of illicit drugs is the 'crash' described after a binge or 'run' of cocaine use, in which there is a sudden fall in mood and an overwhelming feeling of fatigue, which contrast sharply with the elation and energy initially associated with stimulant use. In the laboratory setting of the US National Institute of Drug Abuse, the surprising observation was made that this 'crash' begins while blood cocaine levels are still very high (but consider the similar effects of alcohol on mood described in Chapter 6). More recently, Curran and Travill (1997) have described a mid-week depression which is common 5 days after weekend use of Ecstasy.

After-effects may be associated with other substances; for example, an anxious, irritable hangover may follow short term sedation with barbiturates. I have been unable to unearth the original source material for claims that a period of lethargy follows the end of an episode of using anabolic steroids, but, as more and more athletes are injecting high doses of steroids, it is worth considering that a vicious *cycle* of drug-induced hyperactivity and aggression alternating with periods of passivity and underactivity might develop in such users. Highs and lows in mood and self-esteem, linked to their perceptions of body image, were reported among the 330 steroid users, aged 15 and over, at the Drugs in Sport Clinic & Users Support service in Durham. The users were not just competitive athletes – a young club doorman who needs 'to look big' might feel he has 'lost everything' when he stops injecting steroids (Kossoff 1998).

Cumulative effects

Some drugs produce a clear cumulative poisoning of the brain (Caan 1996a). These include the designer opioid MPTP (Parkinsonian symptoms develop over time with its use) or the proconvulsant β-carbolines present in various plants including the Syrian rue (seizures begin after repeated doses of these harmala alkaloids). The poisonous 'tares' mentioned in the Bible (in 13 Matt.) were once cultivated by some ancient societies, but this plant (*Vicia*) contains an excitotoxin which can result in both seizures and paralysis over time. I used to imagine that in a modern society few citizens would habitually use such obviously hazardous drugs, but now I am less confident of such common sense prevailing.

An increasingly varied range of mental health problems are being reported in people who have used Ecstasy repeatedly (Black 1992; McGuire *et al.* 1994), and in rats and monkeys this drug has consistently produced a severe depletion of the neurotransmitter 5-HT in the brain. Work continuing at the Johns Hopkins University, using positron emission tomography to compare the brains of regular Ecstasy users with other drug users, has now demonstrated 5-HT deficiencies associated with the use of Ecstasy in all regions of the human brain. Yet it still remains fashionable ...

One explanation for this is that Ecstasy users in general consider the benefits of use *to them* far outweigh the theoretical risks – risks they may believe are exaggerated by the biomedical 'vested interests' (Ashenhurst

1996). A similar problem of credibility and drug marketing may be at work in maintaining anabolic steroid use (ISDD 1994) in the face of multiple harmful and irreversible effects of chronic use; for example, growth of the skeleton can be stunted in children and adolescents, young males can experience shrinking testicles and growing breasts and young females can develop a deep voice and facial hair (HEA 1996). But many young people are indoctrinated that *winning* now is everything in a competitive world, whatever the eventual cost.

The route of administering drugs can be important in accumulating physical health problems over time; for example, public health concern is growing (Hall and Babor 2000) over the cumulative effects of smoking cannabis on developing respiratory disease.

The setting of intoxication can also be important. In a police cell, following accumulated cocaine use over a short period, a case of excited delirium may well prove fatal (Spinney 1999).

The more often street drugs are used, the greater the number and variety of hazards reported. In the British 'NTORS' cohort (DoH 1996), 54 per cent of users had needed Accident & Emergency treatment during the previous 2 years. The average age in that cohort was about 29 years and many had developed a severe dependence on drugs. However, younger people may be at increased risk of certain types of harm, as shown by a study of HIV seroconversion in Amsterdam users (Fennema *et al.* 1997). It was the *youngest* users (under 23 years) with the shortest histories of injecting drugs (under 3 years) who had by far the highest HIV-positive rate.

An early onset of drug use combined with any injecting of drugs is associated with numerous increased risks (Dinwiddie *et al.* 1992a), but that does not imply that young people can anticipate the precise nature of these risks. Heroin is sold by weight and usually 'cut' with other powdery adulterants (like talc), and a few years ago one of the major drug dealers on the small island of Malta decided to adulterate his heroin with something really 'heavy': lead. He therefore added lead acetate to a batch which became the supply to a third of all the islands' drug injectors. Even if they had read all the best textbooks on drugs, those poor people would not have guessed that cumulative lead poisoning would be their most likely hazard!

In many British country towns like Cambridge, amphetamines are used much more extensively than heroin. Both physical and psychological problems *are* common with regular use of amphetamine (Pates and Mitchell 1996) but only 11 per cent of their community sample had ever sought help from any drug agency. Robson and Bruce (1997) studied amphetamine users extensively, including that great majority of users who had never contacted a service for help. Typically, they used amphetamine weekly (not daily), 'were not worried at all' about their drug use and exhibited little dependence. The amphetamine users *seen* by drug services had been using this drug for a longer time, they were much more concerned about their problems, were more likely to inject drugs and were more likely to show dependence, than the 'invisible' users in the community. Sadly, if they *did* contact our current types of service, the majority found them to be useless (58 per cent reported 'minimal' or 'no' help at all).

'Fallout' of drug use: secondary casualties

Substance use affects a wider circle of people than the user herself or himself; for example, more than 3,000 of the children telephoning the ChildLine helpline last year described parents who were misusing alcohol. Alcohol appeared to be a contributing factor to half the calls reporting family violence or sexual abuse of these children (ChildLine 1997). Age effects cannot be simply stereotyped; when crack cocaine first appeared in London in the late 1980s, one of the first related calls to our drugs helpline was from an elderly woman terrorized by her adolescent grandson for money to buy his crack. Many possible problems arise from 'other people' using drugs; for example, if unemployed adults dependent on a daily gram of street heroin need to steal two video recorders or their equivalent per day, for an unemployed teenager wanting one hit of 'smack' this translates into stealing someone else's car radio or its equivalent. However, in this section I will focus on three health-related areas of fallout from drugs: *infectious disease, violent crime* and *complications of pregnancy*.

The relationship in the population between drug use and infectious illnesses was a cause for concern long before AIDS appeared in the 1980s; for example, in the 'Swinging Sixties', Rawlins (1969) found that 63 per cent of his female patients presenting with sexually transmitted diseases (gonorrhoea or syphilis) had been taking amphetamines. People who inject cocaine hydrochloride ('snow') are at great risk of HIV infection in the course of repeated unsterile injections (especially within a large group of simultaneous users like an urban 'shooting gallery' or on a prison wing), but it is now clear that smoking the cocaine base ('crack') can also increase the risk of HIV, through changed sexual behaviour (Des Jarlais 1991) and, once a female user is infected, the virus may then be *passed on* to her future offspring. A decade ago in Britain, the government misjudged the effect of their advertisements, after which young users associated HIV infection with dirty 'street' heroin but not with their unsterile injections of pharmaceutically 'pure' alternatives like buprenorphine or temazepam. Remember, it is the *youngest* users without long histories of dependent drug use who are most at risk of acquiring HIV (Fennema *et al.* 1997). These young people have their reproductive lives ahead of them.

Thanks to some media coverage and peer education, most 16-year-olds in the UK have heard of HIV and AIDS. However, the statistical association of drug use with diseases like tuberculosis and hepatitis is much less well known. A quarter of the drug users sampled in the community by Robson and Bruce (1997) already had the hepatitis B virus, so what proportion of *their* partners, children and grandchildren will be HBV-positive? Hepatitis C is another virus readily transmitted intravenously, and, in many environments (like the prisons confining young women: Devlin 1997), this virus has infiltrated an unsuspecting community of adolescent drug users. In one London sample of drug injectors, before the risks of HCV transmission were appreciated, 71 per cent became infected with this virus (Stimson and Hunter 1998).

At last in the UK, domestic violence is recognized as a legitimate concern for health services (Calman 1997). In London, the first two referrals to our drug clinic for acute crack-related problems had both committed serious assaults on their partners. Periodic feelings of suspicion and hostility are pretty common amongst high-dose stimulant users in general, and most users also combine high doses of alcohol with their amphetamine or cocaine binges (see Chapter 6 for alcohol and morbid jealousy). Survivors of rape or childhood sexual abuse may experience adverse mental health effects for many years afterwards. In the course of my drug research in male prisons (Caan 1991), I listened to over sixty sex offenders describing their substance use at different stages of their lives. The majority had led long antisocial careers of acquisitive crime (like burglary or robbery), in the course of which they had also assaulted women. With different offenders, one story kept being described: during runs of combined stimulant and alcohol bingeing, these men committed their most violent acts, usually against strangers they were robbing. The combination of excitable aggression (with cocaine) along with disinhibition and remorselessness (with alcohol) seems to cause an escalation of offending, in the context of committing antisocial acts. Across European society as a whole, 17 is the age at which criminal behaviour is *most common* (Rutter and Smith 1995). The study of US schoolchildren by Kingery *et al.* (1996) found that those adolescents who carried a gun were seventeen times more likely to have used crack cocaine than the other, unarmed, teenagers. In its study of six US cities between 1987 and 1993, the National Institute of Justice found a 'very strong statistical correlation' between the yearly *homicide* rates and the population rate of adult male arrestees who tested *positive for cocaine.* In Britain as early as 1993, the future 'drugs tsar' Keith Hellawell was reporting to the Chief Police Officers conference, against the background of escalating use of both amphetamines and cocaine, that 93 per cent of young offenders were 'regular drug users' (note that only 6 per cent had ever been charged with a *drugs* offence). Many factors influence youth crime, but cannabis and Ecstasy use were prominent in a study of offences like shoplifting around the ages of 12–16 (Patel 1995). In some crimes, it is the victim who is drugged, involuntarily. In the early 1990s, I was asked several times to comment on investigations of the sexual abuse by adults of minors who had been administered hallucinogens (probably LSD). These violent attacks seemed doubly distressing for the teenage victims (see 'bad trips', p. 153) all of whom had difficulty describing the exact time and place of the abuse. Although a number of other sedatives, such as methaqualone or gammahydroxybutyrate, have been reported to make young people more acquiescent about sex with strangers, alcohol has always been the main drug associated with the sexual abuse of minors. A worrying trend picked up in 1997 by Scotland Yard is 'date rape' following administration of a little alcohol 'spiked' with the short-acting benzodiazepine flunitrazepam. These small tablets are the 'roaches' (see Chapter 15) favoured by US benzodiazepine abusers, and have been pretty alien to our UK drug scene which has for long favoured big capsules of temazepam for *self*-administration.

The use of drugs during pregnancy

This is the most emotive of all issues within the already emotive area of drug use and young people. The DoH (1996) list 'possible obstetric complications among pregnant drug users' as one of the 'harms' associated with drug misuse, nationally. With alcohol, it has taken many years of research for a clear picture of problems like Foetal Alcohol Syndrome to emerge (Caan 1996b).

The most widely used illicit drug, cannabis, may have harmful effects on foetal growth and development, as does tobacco, but I shall use examples from the literature on cocaine to illustrate some problems in pregnancy, as there is a wealth of recent research published. The effect best known to the public relates to neonatal withdrawal problems in about 25 per cent of 'crack babies' (Strang *et al.* 1993), but there are also specific hazards during pregnancy like a fourfold increase in the risk of abruptio placentae (Hulse *et al.* 1997a) and, on average, a 112 g lower birth weight at delivery (Hulse *et al.* 1997b). In some parts of the US, draconian punitive reactions to pregnant girls using cocaine have been instituted (e.g. incarceration before and during their delivery), but it is valuable to consider the feelings of the *mother* during and after her pregnancy: in most cases, these mothers have been pregnant before (Kearney 1993), and they do aim to reduce the risk of foetal damage and aim to keep custody of their baby while avoiding the painful stigma inflicted by professionals.

Individuals and groups which may be especially vulnerable

Individuals who have experienced certain childhood difficulties (such as a disrupted home life and conduct problems: Dinwiddie *et al.* 1992b) are more vulnerable to adopting *intravenous* drug use in adolescence. In the US literature, the label 'antisocial personality disorder' is frequently associated with teenage use of illicit drugs, but at a Cambridge conference ('The

Challenge of Personality') the advice of a leading European authority, Peter Tyrer, was that the relationship between specific patterns of drug use and any clinically useful classification of personality disorder has yet to be demonstrated. Some relationship between types of hazardous drug use and personality may very well exist, but our scientific understanding of personality and its maturation needs to be deeper (compare with alcohol, Chapter 6). What is clear is that more and more individual teenagers are taking a variety of substances (Rutter and Smith 1995).

Four *groups* of people are well recognized to be at increased risk of health problems consequent to drug use. *Homeless* women show especially high rates of drug use, in combination with other illnesses, particularly depression (Nyamathi *et al.* 1997). Drug users may be evicted from their homes within the Crime and Disorder Act as 'nuisance neighbours', creating a vicious cycle of homelessness. Suicide in young homeless people of both sexes is a matter of growing concern in the UK. The charity LANCE offered a 'mentoring' scheme to support young people at risk who live on the streets of Cambridge. Half the inner-city residents with *severe mental illness* aged 20–29 also have drug problems (Menezes *et al.* 1996) which appear to aggravate their illness. Caan and Crowe (1994) found that, of all the possible 'dual diagnosis' combinations, a co-morbid diagnosis of dependence on alcohol or drugs predicted the most frequent re-admissions to psychiatric hospitals. A *chronic, intermittently painful, physical* condition (such as sickle cell anaemia) which introduces young patients to powerful painkillers like pethidine may influence their subsequent attitude to seeking opiates. Lastly, *prisoners* are at great risk of harm from drugs (DoH, 1996). Within the prisons, it is possible to make headway in some areas of harm reduction; for example, increasing the rates of immunization for hepatitis B from 2 per cent of inmates to over 80 per cent (Williams *et al.* 1996). What may be more effective still is keeping young drug users out of the prison environment altogether, through appropriate court diversion schemes (McFadyean 1997).

The Good Practice Unit for Young People and Drug Misuse have identified that vulnerable young people often combine several risk factors: 'A child with a history of family problems, for example, might be experiencing problems at school, may have been accommodated (or currently in) local authority care and may also be offending on a regular basis' (SCODA 1997). The UK's national drugs strategy (Howard 1998) will, hopefully, transform a situation where nine out of ten social services departments have had no budget to address such vulnerable young people's drug problems.

Preventable harm

In Chapter 18, the overall features of prevention will be addressed at a national level of planning. In this section, I will briefly sketch the health promotion issues which relate particularly to young people. If you work with children and adolescents and want more detailed guidance, see McEwen and Clancy (1997) who give excellent advice on how to *engage* with them around their (or their parents') drug concerns, especially taking into account the young person's perspective and their significant relationships. If you are involved in planning or co-ordinating care for young people, see the Health Advisory Service (1996) who take into account the wide variation in needs different adolescents may have, for example:

- *direct access, generic services* involving professionals like family doctors, teachers and school nurses;
- *youth-oriented professionals with specialist knowledge* including educational psychologists and social workers from child and family services;
- *specialist multidisciplinary addiction teams for young people* within mental health, obstetric or forensic services for adolescents;
- *intensive forms of intervention* such as residential drug rehabilitation or inpatient psychiatric units for adolescents.

The vast majority of cases should be managed at the first two levels listed above.

Reducing substance use

In planning interventions for young people, it is important to consider both the ways to prevent first drug use in those who have never tried drugs and how to minimize the amount of use in those who are trying drugs already; for example, peer education may have more of an impact on preventing use than reducing established use (Ward *et al.* 1997). The HEA (1997b) recently completed a review of effective methods for mental health promotion, including a wide range of techniques demonstrated to reduce drug use in school age children. All the effective methods involved the active participation of young people (and often their families or teachers as well) and included social skills, problem solving, assertiveness or communication skills training, in addition to education about drugs and health. These preventive approaches could be complemented by cognitive-behavioural training (to build self-esteem, manage anxiety and improve relationships) to reduce pupils' level of drug use. The 'Teenage Health Club', set up by a school nurse (Little 1997) in an inner city area with high levels of drug use, involved teenagers actively running the club and choosing their own topics for health promotion. The model of health was based on developing the self-esteem of the adolescents in an environment including active leisure skills, exercise and group games – teenage girls even invited mothers to enable conjoint drugs education. Two imaginative approaches which have recently undergone promising evaluations are 'Project Charlie' in urban

East London which involves games for 10-year-olds that promote decision-making skills, and 'Drugs on the Streets' in rural Hampshire which involves mobile youth theatre and take-home reminders on T-shirts. The English agency TACADE offers help for teachers and youth workers on many aspects of substance use, including communication skills training or building self-confidence for teenagers, and specialist problems like drug misuse in sport. Compact disk material about drugs, for adolescents themselves to use with a multi-media computer, has been developed by the Health Education Authority with the title DCode.

These multifaceted initiatives are beginning to have a national impact on schoolchildren; for example, the use of cannabis seems now to be falling in adolescents (Hayes 2000).

Minimizing the harm associated with drug use

Needle exchanges are a good example of harm minimization around HIV for drug injectors, but to get the full benefit from direct access services (Caan 1994) harm reduction goals should be *negotiated* individually with drug users, in the context of a growing therapeutic partnership over time between the user and one key worker. Dialogue between services and individual users or user groups is a vital step in promoting *responsibility* in safer drug use (Hall 1997). The Health Advisory Service (1996) recognize the value of a safe environment at raves in which an Ecstasy user could rehydrate and 'chill out' after strenuous dancing. The Mental Health Foundation (1998) involved opiate users in an initiative to prevent the spread of injecting behaviour, building on first-hand experiences of the initiation into injecting drugs. A collaborative initiative in Plymouth between the probation and community health services has reduced criminal behaviour in drug-using offenders who would otherwise be in prison (Crime Prevention 1997). Soon, I hope to see safety programmes around cannabis and driving similar to current 'drink-drive' campaigns, but this is still waiting for political approval.

Resilience

Many scientists have sought to elucidate the factors which make some adolescents especially vulnerable or susceptible to drug problems, since it is obvious that not all young people are affected equally by our current culture of drug use. Recently, there has been a growing interest in factors that might *promote resilience* in young people; for example, in relation to the World Mental Health Day Campaign (1996), which emphasized self-expression and creativity. Religious devotion can function as a buffer for stressful events (Kendler *et al.* 1997) and reduce the risk of dependence developing for either alcohol or nicotine. Four specific attributes of resilient US youth have been identified in high-risk environments for drug problems:

- social competence;
- problem-solving skills;
- autonomy (with a sense of one's identity, independence and control over events);
- a sense of purpose and future.

And Quinn (1995) described the positive effects of active participation in a variety of *youth groups* on all the above features of resilience; for example, participation in the Boys Clubs of America was associated with reduction in drug use, drug trafficking and drug-related crime.

The relationship between the resilience shown by individuals and the 'social capital' built up in their communities has recently become a major focus of interest in the UK. Experiencing 'social trust' in a church or neighbourhood group can be very valuable for a young drug user (Cooper *et al.* 1999). For the younger child, growing up in an environment which offers life chances around education, employment and a social network of friendly neighbours makes a long-term impact on their health behaviours like smoking and drinking. Even pre-school experiences may have a profound influence on later health, as in the national 'Sure Start' programme *enabling* families with children aged 0–4, who live in deprived neighbourhoods, to improve their local social capital.

Rethinking the health and lifestyle contexts which might promote problem drug use

In their review for the Cambridge and Huntingdon Health Authority, Moore *et al.* (1997) began with the premise that efforts to promote adolescent health had to be based on the assessment of young people's need, enable their participation and empowerment, support partnerships across agencies and result in sustainable change. Two practical strands then emerged for health promotion:

- developing life skills with the adolescents;
- involving young people in local community development projects.

A test

Can you think of a life skill *you* could communicate to a 13-year-old? Can you think of a *local* community project to which that 13-year-old could contribute?

Adolescence will always be a time one learns about life; given the right support by familiar caring professionals, it might also become a time to learn about living and about living together.

References

Adams, S. and Silverstone, T. (1991) 'Alcohol and drug abuse prior to first psychiatric admissions', *Brit. J. Addict.*, **86**, 1518.

Allen, D. F. and Jekel, J. F. (1991) *Crack: The Broken Promise*. London: Macmillan.

Anderson, H. R. (1990) 'Increase in deaths from deliberate inhalation of fuel gases and pressurised aerosols', *Brit. Med. J.*, **301**, 41.

Andreasson, S., Allebeck, P., Engstrom, A. and Rydberg, U. (1987) 'Cannabis and schizophrenia', *Lancet*, **ii**, 1483–6.

Ashenhurst, A. (1996) 'The ecstasy sub-culture: Beyond the rave', *Brit. Med. Anthropol. Rev.*, **3**(2), 22–7.

Ashton, C. H. (1987) 'Cannabis: Dangers and possible uses', *Brit. Med. J.*, **294**, 141–2.

Bean, P. (1993) *Cocaine and Crack. Supply and Use*, pp. 90, 94. London: Macmillan.

Black, T. (1992) 'An informed introduction to 3,4 methylenedioxymethamphetamine (MDMA, 'Ecstasy', 'E')', unpublished BSc thesis, University of Hull.

Bloomfield, R. and Kerr, J. (2000) 'Matters of substance', *Time Out*, **1547**, 18–19.

Caan, W. (1991) 'Preliminary observations on a drugs and AIDS initiative for prisoners in the South West Thames health region', *Brit. J. Addict.*, **86**, 1516.

Caan, W. (1994) 'Community psychiatric nursing: Drug-use and HIV-risk characteristics', *J. Psychiat. Mental Health Nurs.*, **1**, 49.

Caan, W. (1996a) 'Exogenous drugs and brain damage', *Addictive Behaviour: Molecules to Mankind*, pp. 56–68. London: Macmillan.

Caan, W. (1996b) 'Foetal alcohol related developmental deficit: A chance for mental health promotion in primary care?', *The Network Developing Mental Health in Primary Care*, **2**, 3.

Caan, W. (1997a) 'Bitter pill', *Nursing Standard*, **12**(3), 21.

Caan, W. (1997b) 'Ecstasy research prompts flashback', *Nursing Standard*, **11**(47), 10–11.

Caan, W. and Crowe, M. (1994) 'Using readmission rates as indicators of outcome in comparing psychiatric services', *J. Mental Health*, **3**, 521–4.

Calman, K. (1994) 'Teenagers taking risks with health at an earlier age', *The Independent* (London), 22 September 1994.

Calman, K. (1997) *On the State of the Public Health 1996*. London: Her Majesty's Stationery Office.

Cawley, S., Praveen, S. and Salib, E. (1997) 'Brief psychiatric admissions: A Review', *Nursing Standard*, **12**(10), 35.

ChildLine (1997) 'News. Children report drunk parents', *Community Care*, 13 November 1997, 7.

Coleridge, J., Cameron, P. A., Drummer, O. H. and McNeil, J. J. (1992) 'Survey of drug-related deaths in Victoria', *Med. J. Australia*, **157**, 459–62.

Commission of the European Communities (1995) 'A summary of the European Commission's proposal to the European Parliament and Council: Action plan to combat drugs (1995–2000)', *Drugs: Education, Prevention and Policy*, **2**, 17–20.

Connoll, P. (1958) 'Amphetamine psychosis', *Maudsley Monograph*, **5**.

Cooper, H., Arber, S., Fee, L. and Ginn, J. (1999) *The Influence of Social Support and Social Capital on Health*. London: Health Education Authority.

Corkery, J. M. (2000) 'Snowed under – is it the real thing?', *Drug Link*, May/June, 12–15.

Crime Prevention (1997) 'Clean break', *Guardian Society*, 23 July 1997, 2–3.

Curran, H. V. and Travill, R. A. (1997) 'Mood and cognitive effects of 3,4-methylenedioxymethamphetamine (MDMA, 'ecstasy'): Week-end 'high' followed by mid-week low', *Addiction*, **92**, 821–31.

Demos (1997) *Young People and Drugs*. York: Joseph Rowntree Foundation.

DoH (1996) *The Task Force to Review Services for Drug Misusers*. London: Department of Health.

Deschamps, A. (1932) *Ether, Cocaine, Hachisch, Peyotl et Demence Precoce*. Paris: Vega.

Des Jarlais, D. (1991) 'Cocaine and the AIDS problem', paper given at Cocaine as a Biological and Medical Problem, CIBA Foundation symposium, 23 July 1991, London.

Devlin, A. (1997) 'Prisoner: Cell Block Hell', *Guardian*, 19 August 1997, 5.

Di Fabio, R. P. and Seay, R. (1997) 'Use of the "fast evaluation of mobility, balance, and fear" in elderly community dwellers: Validity and reliability', *Physical Therapy*, **77**, 904–17.

Dinwiddie, S. H., Reich, T. and Cloninger, C. R. (1992a) 'Lifetime complications of drug use in intravenous drug users', *J. Substance Abuse*, **4**, 13–18.

Dinwiddie, S. H., Reich, T. and Cloninger, C. R. (1992b) 'Prediction of intravenous drug use', *Comprehens. Psychiat.*, **33**, 173–9.

Drugs Misuse Team (2001) *Government Action Plan to Prevent Drug-Related Deaths*. London: Department of Health.

DTI (1990) *A Review of Research on Falls among Elderly People*. London: Department of Trade and Industry.

Fennema, J. S. A., van Ameijden, E. J. C., van den Hoek, A. and Coutinho, R. A. (1997) 'Young and recent-onset injecting drug users are at higher risk for HIV', *Addiction*, **92**, 1457–65.

Hall, S. (1997) 'An injection of hope', *Community Care*, 27 November 1997, 26–27.

Hall, W. and Babor, T. F. (2000) 'Cannabis use and public health: Assessing the burden', *Addiction*, **95**, 485–90.

Hawton, K., Fagg, J., Simkin, S., Bale, E. and Bond, A. (1997) 'Trends in deliberate self-harm in Oxford, 1985–1995', *Brit. J. Psychiat.*, **171**, 556–60.

Hayes, G. (2000) 'School age drug use falling', *Drug Link*, May/June 2000, 7.

HEA (1996) *A Parents Guide to Drugs and Solvents*. London: Health Education Authority.

HEA (1997a) *Health in England 1996*. London: Health Education Authority/ ONS.

HEA (1997b) *Mental Health Promotion. A Quality Framework*. London: Health Education Authority.

Health Advisory Service (1996) *Children and Young People: Substance Misuse Services*. London: Her Majesty's Stationery Office.

Hermle, L., Gouzoulis, E., Oepen, G., Spitzer, M., Kovar, K. A., Borchardt, D., Funfgeld, M. and Berger, M. (1993) 'Zur Bedeutung der historischen und aktuellen Halluzinogenforschung in der Psychiatrie', *Nervenarzt*, **64**, 562–71.

Howard, R. (1998) 'Hooked on hope?', *Community Care*, 8 October 1998, 18–24.

Hulse, G. K., Milne, E., English, D. R. and Holman, C. D. J. (1997a) 'Assessing the relationship between maternal cocaine use and abruptio placentae', *Addiction*, **92**, 1547–1551.

Hulse, G. K., English, D. R., Milne, E., Holman, C. D. J. and Bower, C. I. (1997b) 'Maternal cocaine use and low birth weight new borns: A meta-analysis', *Addiction*, **92**, 1561–70.

ISDD (1994) *Drug Abuse Briefing*, **5**, 37–8.

Jansen, K. L. R. (1993) 'Non-medical use of ketamine', *Brit. Med. J.*, **306**, 601–2.

Kearney, M. H. (1993) 'Salvaging self: A grounded theory study of pregnancy on crack cocaine', unpublished PhD thesis, University of California.

Kendler, K. S., Gardner, C. O. and Prescott, C. A. (1997) 'Religion, psychopathology, and substance use and abuse: A multimeasure, genetic-epidemiologic study', *Amer. J. Psychiat.*, **154**, 322–9.

Kingery, P. M., Pruitt, B. E. and Heuberger, G. (1996) 'A profile of rural Texas adolescents who carry handguns to school', *J. School Health*, **66**, 18–22.

Kossoff, J. (1998) 'Steroid abuse spreads to clubs and bars', *The Independent* (London), 27 September 1998, 13.

Kraeplin, E. and Lange, J. (1927) *Klinische Psychiatrie*. Leipzig: Barth.

Leon, D. A., Chenet, L., Shkolnikov, V. M., Zakharov, S., Shapiro, J., Rakhmanova, G., Vassin, S. and McKee, M. (1997) 'Huge variation in Russian mortality rates 1984–94: Artefact, alcohol, or what?', *Lancet*, **350**, 383–8.

Lewin, L. (1924) *Phantastica*. Berlin: Stilke.

Lewinsohn, P. M., Rohde, P. and Brown, R. A. (1999) 'Level of current and past adolescent cigarette smoking as predictors of future substance use disorders in young adulthood', *Addiction*, **94**, 913–21.

Linzsen, D. H., Dingemans, P. M. and Lenior, M. E. (1994) 'Cannabis abuse and the course of recent-onset schizophrenic disorders', *Arch. Gen. Psychiat.*, **51**, 273–9.

Little, L. (1997) 'Teenage health education: A public health approach', *Nurs. Standard*, **11**(49), 43–6.

Lord, S. R., Anstey, K. J., Williams, P. and Ward, J. A. (1995) 'Psychoactive medication use, sensori-motor function and falls in older women', *Brit. J. Clin. Pharmacol.*, **39**, 227–34.

McEwen, A. and Clancy, C. (1997) *Substance Use: Guidance on Good Clinical practice for Nurses, Midwives and Health Visitors. Working with Children & Young People*. London: ANSA.

McFadyean, M. (1997) 'Out of the drug trap', *Guardian*, 9 December 1997, 17.

McGrath, M. (1997) 'Fear and pudding in Vegas', *The Big Issue*, 17 November 1997, 8–9.

McGuire, P. K., Cope, H. and Fahy, T. (1994) 'Diversity of psychopathology associated with use of 3,4-methylenedioxymethamphetamine ('Ecstasy')', *Brit. J. Psychiat.*, **165**, 391–5.

Maier, H. W. (1926) *Der Cocainismus, Geschichte, Pathologie, medizinische und behordliche Bekampfung*. Leipzig: Thieme.

Marks, H. (1997) 'The great supergrass debate', *The Big Issue*, 17 November 1997, 14.

Mathers, D. C., Ghodse, A. H., Caan, A. W. and Scott, S. A. (1991) 'Cannabis use in a large sample of acute psychiatric admissions', *Brit. J. Addict.*, **86**, 779–84.

Menezes, P. R., Johnson, S., Thornicroft, G., Marshall, J., Prosser, D., Bebbington, P. and Kuipers, E. (1996) 'Drug and alcohol problems among individuals with severe mental illnesses in South London', *Brit. J. Psychiat.*, **168**, 612–19.

Mental Health Foundation (1998) 'Preventing initiation into injected drug use', *MHF Briefing*, **20**.

Miller, P. M. and Plant, M. (1996) 'Drinking, smoking and illicit drug use among 15 and 16 year olds in the United Kingdom', *Brit. Med. J.*, **313**, 394–7.

Moore, V., O'Meara, I., Hanby, G. and Burton, H. (1997) *Young People's Health Promotion Framework*. Cambridge: Cambridge & Huntingdon Health Authority.

NCH Action for Children (1997). *The Age of Anxiety*. London: NCH.

Nyamathi, A., Flaskerud, J. and Leake, B. (1997) 'HIV-risk behaviours and mental health characteristics among homeless or drug-recovering women and their closest sources of social support', *Nurs. Res.*, **46**, 133–7.

Parkman, S. (2000) 'Mental health', *Community Care*, 27 April 2000, 27.

Patel, K. (1995) 'Drugs and crime link', *Times Higher Education Supplement*, 27 January 1995, 7.

Pates, R. M. and Mitchell, A. (1996) 'Amphetamine use', *J. Substance Misuse for Nursing, Health and Social Care*, **1**, 165–73.

Petersen, R. C. (1978) 'The international challenge of drug abuse', *NIDA Research Monograph*, **19**.

Phillips, P. and Johnson, F. (2000) 'Preventing heroin overdose', *Nurs. Standard*, **14**(24), 31.

Poole, R. and Brabbins, C. (1996) 'Drug induced psychosis', *Brit. J. Psychiat.*, **168**, 135–8.

Quinn, J. (1995) 'Positive effects of participation in youth organizations', *Psychosocial Disturbances in Young People. Challenges for Prevention*, pp. 274–304. Cambridge: Cambridge University Press.

Rawlins, D. C. (1969) 'Drug-taking by patients with venereal disease', *Brit. J. Venereal Dis.*, **45**, 238–40.

Release (1997) *Drugs and Dance Survey*. London: Release.

Roberts, C., Moore, L. Blakey, V. Playle, R. and Tudor-Smith, C. (1995) 'Drug use among 15–16 year olds in Wales, 1990–94', *Drugs: Education, Prevention and Policy*, **2**, 305–16.

Roberts, I., Barker, M. and Li, L. (1997) 'Analysis of trends in deaths from accidental drug poisoning in teenagers, 1985–95', *Brit. Med. J.*, **315**, 289.

Robson, P. and Bruce, M. (1997) 'A comparison of 'visible' and 'invisible' users of amphetamine, cocaine and heroin: Two distinct populations?', *Addiction*, **92**, 1729–36.

Rutter, M. and Smith, D. J. (1995) *Psychosocial Disorders in Young People*. Chichester: Wiley.

SCODA (1997) *Drug-related Early Intervention*. London: SCODA.

Smith, D. (1998) 'Marijuana: Drop in with Dr. Dave', *New Scientist*, 21 February 1998, 26.

Spinney, L. (1999) 'Tragedy behind bars', *New Scientist*, 25 September 1999, 4.

Stimson, G. V. and Hunter, G. (1998) 'Public health indicators', in G. V. Stimson, C. Fitch and A. Judd (eds) *Drug Use in London*, pp. 198–214. London: Leighton Print.

Strang, J., Johns, A. and Caan, W. (1993) 'Cocaine in the UK – 1991', *Brit. J. Psychiat.*, **162**, 1–13.

Swain, H. (1997) 'Plea for a war on drugs', *The Times Higher*, 27 June 1997, 4.

Ward, J., Hunter, G. and Power, R. (1997) 'Peer education as a means of drug prevention and education among young people: An evaluation. *Health Educ. J.*, **56**, 251–63.

Watts, D. (1997) 'Brief encounters', *Nurs. Times*, **93**(30), 28–9.

Williams, O., Cassidy, J. and Caan, W. (1996) 'Banged up, shooting up, messed up', *Health Service J.*, 27 June 1996, 26–7.

Wolff, P. O. (1945) *The Treatment of Drug Addicts*, Bulletin of the Health Organisation, Vol. 12, No. 4. Geneva: League of Nations.

World Mental Health Day Campaign (1996) 'Positive images, positive steps', *Young People's Health Network*, **1**, 3.

Zimmern, R. (1997) *Chance & Choice ... Balancing the Risk*. Cambridge: Cambridge & Huntingdon Health Authority.

Chapter 14

The nature of heroin and cocaine dependence

Woody Caan

Key points

- *The availability of heroin and cocaine has been rising dramatically, worldwide.*
- *Early experiences of using either drug (the opiate or the psychomotor stimulant) are often quite positive and if, subsequently, dependent use develops this usually grows over a period of years.*
- *The two drugs act on quite different neuronal targets: opiates bind to synaptic receptors for the peptides called 'enkephalins'; cocaine blocks the monoamine transporters necessary for the removal of dopamine from synapses.*
- *The brain systems involved are similar, including regions of the pre-frontal cortex and basal ganglia with important functions in mood and memory.*
- *Once an 'addiction' to either drug has become entrenched, an individual's life is likely to show a pattern of disorder which combines biological, psychological and social changes.*
- *The pattern and route of drug use, the social context of using and the place of drug use within the life story of the user have major effects on the evolution of drug dependence.*
- *There is NO stereotypical 'addict' – men and women in many walks of life and in all ethnic and political groups can develop dependence on these powerful substances. Conversely, recovery is seen in many walks of life, as countless women and men win their independence from addiction.*
- *During dependence, social handicap and physical morbidity can place a great burden on users of both substances.*
- *Drug seeking is prominent for both drugs, but they often differ in the pattern produced.*
- *Regular heroin users frequently take opiates to avoid physical withdrawal symptoms when they have been deprived of their drug (e.g.*

> *overnight), whereas this is a minor problem with cocaine use. Using a little cocaine can stimulate insatiable cravings for more, so that a cycle of accelerating use and exhausted crashes is a typical feature of cocaine dependence, whereas some opiate users maintain fairly stable daily doses determined by their long-term level of tolerance.*
>
> - *While both drugs produce psychological changes, the patterns of thought learned under the influence of cocaine can persist, even during years of abstinence.*
> - *In the past, drug treatment with a narrow focus on biological problems demonstrated a mismatch with patients' aspirations, which usually focus on urgent social needs.*
> - *Now that the range of client needs is becoming understood, multi-professional teamwork is being established in a range of treatment settings, and better and better outcomes are being reported.*

Dependence on illicit drugs is quite common: the yearly rate in the UK is 3 per cent of adult males and $1\frac{1}{2}$ per cent of adult females, with a sharp peak for men aged 20–24 (OPCS 1994). The United Nations have identified trade in heroin and cocaine as the largest value *international* commerce on our planet (it has overtaken armaments or oil since the end of the Cold War). We are talking about a problem for *humanity*, and very common, very human attributes can contribute either to developing dependence or to recovery. However, to people who have never experienced addiction, it can appear a very puzzling phenomenon.

The pursuit of happiness and the human brain

The predominant components of 'happiness' have been the focus of research by Argyle (1992) for many years. His resultant Oxford Happiness Inventory has three dimensions:

- the frequency of experience of *joy* and its *intensity*;
- level of *satisfaction* with life as a whole; and
- the *absence* of depression, anxiety or other *negative* states.

These are also the dimensions of reinforcement which maintain and entrench the use of both heroin and cocaine.

Heroin

Heroin works initially on all these three dimensions, as do the related products of the opium poppy like morphine or the synthetic opioids like methadone, pethidine, dipipanone, buprenorphine and fentanyl. During the first period of use (long before dependence has developed), smoking or injecting heroin gives a rush of pleasure (the 'hit' soon after each use) which is very intense and preoccupying. Over the first year or so of frequent 'recreational' use, tolerance develops to this action of opiates and the joyful rush requires higher and higher doses of drug to recapture, eventually fading out even with a five to tenfold escalation of the dosage, say from adulterated heroin equivalent to 10 mg pharmaceutical diamorphine up to single injections containing over 50 mg (a 'quarter gram bag' of street heroin in one shot). In a beginner, heroin also produces a prolonged feeling of contentment with life, perhaps 4 hours of satisfaction after inhaling the smoke from a small dose heated up on tinfoil with a cigarette lighter ('chasing the dragon').

Tolerance to this positive effect develops more gradually, but, by the time users seek help for drug dependence, most gain only a very, very brief interval of contentment following heavy doses. The heaviest *sustained* level of use I have met so far (a habit built up in London and Asia over 20 years) was 2.25 g daily, equivalent to perhaps 700 mg diamorphine – more than enough to kill any non-addict, even if divided into five injections per day – but this woman had not experienced any joy or satisfaction for years. The remaining effect of heroin is the most *dependable*, across time and across users: a feeling of *safety*, of being 'wrapped in cotton wool', *immune* from life's pains, miseries and humiliations. As physical dependence on opiates develops, withdrawal effects include negative mental states (anxiety, craving) and distressing physical problems (diarrhoea, nausea, sweats, cramps, insomnia). So, just like the alcoholic seeking to drink an 'eye-opener' (see Chapter 5) on waking, heroin addicts begin taking their drug to relieve withdrawal symptoms. The Opiate Withdrawal Symptom Questionnaire (Ghodse 1989) provides a useful tool to check the profile of heroin withdrawal problems. Most of our clinic patients have been using heroin for a decade or more, and the *avoidance of negative states* plays a big part in maintaining their drug use.

Cocaine

Cocaine (like all stimulants) is more variable in its effects on different individuals – individual psychological traits may determine which effects become prominent. Looking at extreme examples, a person with a developmental Attention Deficit Disorder may self-medicate to feel more alert and 'in focus', while, in a person with a pre-existing Obsessive Compulsive Disorder, the drug may amplify their tendency for idiosyncratic, ritual behaviours. The homeless teenager I met recently became loquacious in telling a stranger about his grand schemes for an exciting life, while intoxicated. I could spot that he was an inexperienced user, because this 'high' persisted for about half an hour: with regular use, cocaine gives only a few minutes lift to mood. Davidson *et al.* (1993) have shown in young people that the initial effects which predict a second use of cocaine at the earliest time after initiation are: *Cocaine made me feel like I could do anything* and *Cocaine made me feel as though I was on top of things*. Early effects which predicted the subsequent frequency of use included positive effects like *Cocaine made anything I talked about more interesting* but also one important negative effect *I was never satisfied when I was on cocaine ... I always wanted more*.

In general terms, for people at risk of future dependence, cocaine makes enjoyable experiences more intense, and, because the effects are short lived (especially if the drug is smoked or injected to give an immediate 'rush': Strang *et al.* 1993), users can quickly learn a pattern of frequently repeated use, building up to, say, ten to twenty doses in a day.

Naive users who inject about 10 mg or sniff about 30 mg of pharmaceutical cocaine produce a distinct effect on their mood, but regular users might take multiple 100 mg doses totalling 2 g in a day with only mild and brief pleasure, due to the development of tolerance. At present, the cocaine sold either as hydrochloride powder or rocks of freebase is much purer than a decade ago – street cocaine which once was about 50 per cent adulterants is now often 90 per cent pure. The heaviest user I encountered in London was a woman in her early 20s who had chain-smoked £2,400 worth of crack in the week before admission to hospital, a binge consuming about 24 g of cocaine in almost 100 rocks. Mood sinks lower and lower in between runs and, eventually, users are struggling to keep their spirits up cyclically with the stimulant, like a gambler 'chasing his losses' with bigger and bigger bets (Strang *et al.* 1993).

Neurobiology: some of the brain systems affected

The common *biological* message in addiction is that the brain is 'a self organising system' (McCrone 1997) in which neurochemistry affects behaviour and behaviour affects neurochemistry. Repeated within the structure of all our brains, there operates a framework of identifiable subsystems, but these systems function in astonishingly flexible ways, adapting all the time to fluctuations in our 'internal' world and to the external environment in which we operate. The monamine pathways play an important role in this interaction of brain and behaviour (MRC 1994), especially the dopaminergic pathway originating in the ventral tegmental area, the 5-HT pathway originating in the nucleus raphe and the noradrenergic pathway from the locus coeruleus, all of which can tune and retune the symphony of neuronal activity ranging across wide areas of the brain. Enkephalins act mainly through short-range connections, but these are often at strategic locations like the locus coeruleus. In rats which have been 'shaped' to self-administer *either* cocaine or heroin, a common structure in the basal ganglia, the ventral pallidum, seems to mediate this behaviour (Hubner and Koob

1990). In the human brain, we are beginning to unravel the subtle role of the prefrontal cortex in emotional learning (Damasio 1997).

Neurobiology: at the frontiers of understanding

As described in Chapter 11, opiates bind to the μ-enkephalin receptor on neurones. However, the interaction of activity at many different receptors in the brain is necessary for the *behavioural* effects of opiates to appear. Murtra *et al.* (2000) have demonstrated that, in mice lacking a single gene, the normal behavioural response to morphine as 'rewarding' is lost, although these mice still respond normally to other rewards like food. The gene which was 'knocked out' in these mice is required for the production of one receptor, for the peptide Substance P – in the basal ganglia. The implications for human use of opiates, of blocking the action of Substance P, are unknown.

While it looks as though one neuronal event might be critical in linking animal behaviour to the administration of opiates, the actions of cocaine on brain metabolism appear ever more complex. While the immediate action of cocaine is to block specifically the re-uptake of dopamine released at nerve terminals, there are also changes in mRNA in the basal ganglia called Cocaine- and Amphetamine-Related Transcript (CART). CART has *multiple* effects on rewards (e.g. stopping hungry animals eating food: Chemicon 1999).

Dependence

The World Health Organization has adopted a 'syndrome' approach to defining the presence of 'dependence' (see Chapter 5). The WHO describe seven dimensions of problem in dependence which tend to arise together and to develop together over time:

- a subjective awareness of compulsion to use a drug or drugs, usually during attempts to stop or moderate drug use;
- a desire to stop drug use in the face of continued use;
- a relatively stereotyped drug-taking habit (i.e. a narrowing in the repertoire of drug-taking behaviour);
- evidence of neuroadaptation (tolerance and withdrawal symptoms);
- use of the drug to relieve or avoid withdrawal symptoms;
- the salience of drug-seeking behaviour relative to other important priorities;
- a rapid reinstatement of the syndrome after a period of abstinence.

For clinicians, Gerada and Ashworth (1997) consider that 'the key feature is a *compulsion* to use drugs, which results in *overwhelming* priority being given to drug-seeking behaviour'. Lay people may express the essence of dependence differently, as when Marianne Faithfull (1994) declared:

> No one understands that when you are in this condition, when you are being asked to stop using drugs or die, you can also choose to die.

The psychologist Davies (1997) attacked the biomedical model above, which describes dependence mainly in terms of helplessness, as 'The Myth of Addiction', claiming rather that rational decision making by free individuals is involved (even if other people might see these decisions as 'ill-advised').

> *In practice, drug users arrive at treatment with a variety of different needs and expectations of help, and so it is useful for the professional to have recourse to more than one theoretical 'model' of dependence to engage with this real-life variety of problems and opportunities.*
>
> (Farmer 1997)

Jellinek (1942), who pioneered the systematic description of alcoholism, recognized parallels between some alcoholics and 'cocainists'. The useful history of the Addiction Severity Index (McLellan *et al.* 1980) for rating problems with either heroin, cocaine or alcohol shows that dependence on all these substances has a lot in common. Genetic studies (Handelsman *et al.* 1993) are beginning to suggest that some families share a risk for addiction, involving a range of substances. Some individuals show dependence on *both* cocaine and heroin, or on each *at different times* in their life. However, in the rest of this chapter, I am also going to highlight some subtle differences between dependence on heroin and cocaine, because I believe these have implications for the flexible way we have to address the problems of different 'addicts'.

Heroin: the horse you have to ride without reins

From a 'friend in need' to a constant companion

Diacetyl morphine (heroin) has now been used for 100 years, having first been produced by Bayer for medical use in 1898. For perhaps 4,000 years, opium had been valued for its pain-reducing properties, but, at particular moments in history, such as the coincidence of the trauma of the American Civil War, the introduction of the hypodermic syringe and freely available supplies of morphine, the widespread development of dependence on opiates had been clearly recognized (Snyder and Lader 1988). More recently, the *context* within which people become dependent has been extensively studied in relation to the Vietnam War, where many American servicemen developed heroin problems in the setting of the war, which then ended on returning home to the USA – but relapsed if they returned to Vietnam for further active service. What is less well known is the contemporary situation which can arise when oil industry workers alternate many times per year between remote drilling rigs (where the heroin habit may thrive) and their family home (where heroin use becomes innappropriate).

The best short summary I could find of the perceived *benefits* of starting to use heroin is from 'Lena's story' in Blake and Stephens (1987):

> heroin, which made her feel warm and secure, as if nothing mattered ... all the questions which had haunted her melted away ... 'I felt adequate, that's the best way to describe it, and most of the time I'd felt inadequate ... when I used heroin I suddenly felt I could cope.'

At the time of that interview, after a long history of using, Lena had been drug-free for 2 years. Blake and Stephens observed, though, that 'Heroin is no longer in her body, but *it is ever-present in her mind*'.

Heroin use is increasing steadily in Britain – in Cambridge, police seizures more than doubled in 2000, and the *Drug Misuse Statistics* bulletin confirms a national pattern of increasing demand for help from drug services, from a younger and younger population of drug users (Government Statistical Service 1998). In a 1998 survey by Addaction of 500 drug-taking clients aged over 14, the average age of starting was 15, the users recognized 'problems' from 17, but only began seeking help for addiction at 22, 7 years after their initial use.

So what is happening over those 7 years? Heroin does not burst upon an indifferent youth population: the British Youth Council have identified that drugs are a major concern among adolescents. The recent death in the city of Glasgow of a boy of only 13 from a heroin overdose has raised adult awareness there of a poly-substance culture. This includes complex patterns of temazepam, Ecstasy and opiate use by many young people (Clark 1998). The Schools Health Education Unit have found that *rural* children aged 14–15 are actually more likely to dabble in a variety of illicit substances than city kids (Wilson and Campbell 1998) and 'smack' (heroin) was the drug which they could then experience as 'problematic'. In the words of one contemporary 14-year-old, 'they are probably not good for your body, but who cares?' (Stanley and Godsmark 1997).

Professionals are sometimes perplexed by apparently chaotic but persistent drug use in children – in 1998, the Director of Social Services in Leeds found it 'extremely disturbing' to take a 10-year-old boy into care for his cocaine habit and his 'abysmal' attendance at school. Through his involvement in the voluntary sector, the Prince of Wales (1998) accurately publicized the interaction between social exclusion and drug use in childhood, leading to such severe problems as homelessness. The Institute for the Study of Drug Dependence has now advocated concentrating drug *prevention* activity on 'vulnerable' groups like young people in social services care, those excluded from school, young offenders and the young homeless (Bradley and Mounteney 1998). But social workers are regularly caring for 15- and 16-year-olds who are *already* immersed in heroin use regardless of the 'risks', within a 'no hope culture' (e.g. in the run-down mining towns: Mapp 1998). To develop appropriate services for such adolescent drug users, the Social Services Inspectorate (1997) found the following a critical factor for success:

> The value of involving young people in planning services.

Even involving a few 15-year-olds, like the sixteen care home respondents of Guirguis and Vostanis (1998), can reveal a diversity of drug histories. In this sample, three children in care had initially negative experiences (and stopped using drugs), three continued to smoke cannabis, eight were repeated users 'of more than one drug without a consistent pattern' and two were judged by psychiatrists to be developing a dependency on heroin.

Another critical success factor in relating to current drug users has been summarized by ANSA (1997) as the systemic issues and the significant others:

> an understanding of the important people ... in their lives is important as the client may not appreciate the influences of friends and families, nor the consequent effects of their drug and alcohol use.

As heroin users reach their 20s, a common picture is of 'chaotic lives' (Bean and Winterburn 1997; Adams, 1998). Haarhoff and London (1995) asked a community sample of such opiate users if they had any 'good friends', but 57 per cent had only one or none at all. The insidious evolution of dependence can often be picked up in disturbed social activities and a worsening of mood; for example, the Composite International Diagnostic Interview (Cottler and Compton 1993) includes this question:

> Has there ever been 2 weeks or longer when you lost all interest in things like work or hobbies or things you usually liked to do for fun?

Inducing a mood of depression can itself come to trigger a craving for opiates (Childress *et al.* 1994). However, the reader should not assume that dependence on opiates only develops in some stereotypical 'junkie', a socially isolated, unemployed, self-defeating misfit. A heroin habit can develop among an 'elite' group like fashion models (Frankel 1998). Many health professionals develop addictions, and a 'professional pride which makes them feel immune' contributes to the growth of such a dependence (Coombs 1996). Currently, two-thirds of all the doctors referred for scrutiny by the General Medical Council have developed health problems with drugs or alcohol (BMJ 1998).

When the habit has really taken hold

Over years, users can build up an intimate 'relationship' with 'their' drug. This relationship can exclude other attachments – the addict becomes unavailable for their family or friends (Kroll 1997) who may feel bereaved by a loss from which heroin shields the user themselves. Willis (1977) first questioned the common professional assumption that drugs do not alter the personality, and Ghodse (1989) then found that information from relatives and friends about a patient's personality *before* drug taking started may be very helpful in assessing dependence. The *strength* with which the attachment to heroin can develop is worth illustrating.

After a police round-up ...

One evening in the 1980s, the Metropolitan Police conduct a periodic 'drugs swoop' on the South Bank of the Thames, where drug dealing in the car parks is felt to detract from the cultural and tourist activities of the Capital. Nine habitual opiate users who were expecting to 'score' their drug there are seized together and bundled into one police cell to wait for a court appearance next morning. My friend, who has AIDS, is among those arrested. Searching and supervision of these anxious prisoners is only cursory. My friend has been detained often in the past, and has managed to conceal his injecting 'works' and some 'gear' (the opioid dipipanone) from the police. In the night, faced with a very unpleasant situation and a threatening morning ahead, he naturally injects some of the drug to render himself invulnerable to these negative feelings. While injecting himself, he carefully explains to the others in custody that he has the lethal infectious illness AIDS, so he cannot risk sharing his works with them. Everybody has heard about AIDS. But the other prisoners are also struggling with the same negative emotions he was feeling, so all eight beat up my friend, grab his syringe and the remaining dipipanone and all eight immediately 'fix' up with the same contaminated works.

Real addiction is much stronger than the fear of death. The majority of Londoners dependent on impure street heroin have accidently overdosed and required active resuscitation at some time (Abbasi 1998), but this does not stop them taking heroin again. A few years after the Islamic revolution in Iran, I was discussing drugs policy with the Minister of Health. The revolutionary government had been holding large numbers of public executions of heroin users (not just traffickers) around the country to demonstrate their 'zero tolerance' of this drug in the new society. But the problem of drug dependence had still not gone away ...

Of mice and men ...

Animals like mice and rats can develop a dependence on drugs. This is most readily studied through drug-seeking behaviours in animals (Markou *et al.* 1993). Drugs like heroin and cocaine were shown long ago to lower the threshold for electrical 'brain stimulation reward' in animals, a rather fundamental action on the memory system for organizing behaviour to gain reinforcement. In line with this drug action, it is possible to observe animals seeking those environmental stimuli which they have previously learned to associate with their drug habit – these environmental cues become secondary reinforcers of behaviour in their own right. This nostalgic craving for drug-*related* mementoes can survive, after withdrawing the supply of the drug itself, for a long time in animals. I will return to this sort of 'unforgettable' learning later in relation to human cocaine users. For those readers who do not have a background in the biological sciences, it may already have come as a surprise to discover that animal models have contributed to our understanding of the compulsive, preoccupying nature of dependence, but almost all readers may be surprised to learn that studies of animal behaviour can even contribute to our understanding of drugs in relation to social factors. Benton (1988) had found that giving opiates altered a specific group of social behaviours in mice, such as isolated pups calling for their mothers. Work in Nottingham by Charles Marsden shows that rats that have experienced social exclusion in early life *subsequently* have an altered response to drugs of dependence, which involves specific changes in the neurotransmitters dopamine and 5-HT.

What do these drugs cost society?

Just in Britain the economic cost is at least £4,000,000,000 (more than six billion US dollars) including drug-related crime and health costs, according to the UK Anti-Drugs Coordinator (Hellawell 1998). This economic burden has more than doubled over the last 4 years; for example, in 1999, the National Association for the Care and Resettlement of Offenders found that across all the people arrested by the Cambridge police, 68 per cent now test positive for illegal drugs. Nationally, 51 per cent of men awaiting trial in prison are *dependent* on these drugs (Hirst 1999). Involvement in drug treatment significantly reduces both criminal activity and health problems among heroin users (NTORS 1996). In a Swedish study, any engagement in a treatment programme has dramatically reduced the deaths among heroin addicts, especially for those heroin users who have not yet been infected with HIV (Fugelstad *et al.* 1997). However, health care and criminal justice activities by themselves will never adequately address the complex interaction of heroin and society – a hard-learned lesson in the USA from the

Office of National Drug Control Policy is that *community development* work is also essential for reducing the burden of heroin (Jurith 1998).

Desire without pleasure

One of the recent breakthroughs in our concept of dependence is that the *wanting*, the craving for a drug, does not depend on it producing pleasure (Concar 1998). Very dilute doses of heroin, much too small to give any hedonistic 'hit', can still give rise to sustained drug-seeking activity by an addict. In dependence, those brain structures which are conventionally described in terms of *reward* or *pleasure* develop a chronic hypersensitivity to heroin. The result of this in an addict is not to produce a state of permanent pleasure but of long-lasting *desire*.

This aspect of dependence probably takes longer to develop than tolerance. Some of the mechanisms of tolerance seem to occur at the brainstem level of the μ-enkephalin receptors on monoaminergic cell bodies; for example, in the locus coeruleus, there is an uncoupling of morphine binding sites on the cell surface from the G-proteins which regulate intracellular metabolism. Craving may involve metabolic changes in quite different areas of the brain, the forebrain targets of the ascending monoamine pathways. Brain scans suggest that in human addicts these hypersensitive targets are the amygdala and the anterior cingulate cortex (Concar 1998).

'Scoring' drugs

The main sources of supply for dependent users tend to be *regular dealers* or a *close friend*. Since addicts who are simultaneously users *and* dealers are common and any close friends left to an addict are likely to share their habit, the boundary between these two principal sources of supply is blurred. Ensuring regular scoring (obtaining drugs) is what structures the time and social contacts of the user, whoever their customary supplier is. In economic terms, heroin demand is 'inelastic' (Petry and Bickel 1998), that is, users faced with an increase in price or a reduction in income will give priority to buying heroin over other expenditure, including other drugs like cannabis. In 2000, a research study of police arrestees in five cities who were tested for use of a range of drugs found that 18 per cent were taking heroin and 10 per cent were taking cocaine. This hard drug use required an illegal income of £10,000 to £20,000 per year, mainly from property crime. (Note that the really high-dose users I described at the beginning of this chapter would need even greater incomes.) The Centre for Analysis of Social Exclusion in London has shown that men who had *any* teenage use of hard drugs subsequently earned 'drastically lower' legitimate incomes in their 20s compared with men with no history of hard drugs. Such a young person, with a heroin habit growing over a decade, is left with few alternatives to frequent crime to pay off his dealers. A distinct minority of dependent users get their supply of drugs from a sexual partner, but only a few dependent users will rely on a supply from casual acquaintances in a pub or club.

Beliefs about heroin and the capacity to engage with help

Working in prisons or in services for HIV/AIDS, we have seen quite a few drug-dependent people who had never been in contact with drug treatment services (Williams *et al.* 1996; Caan 1994). On average, compared with the clients attending a drug dependency unit, such 'invisible' users might be a little younger or be a little more likely to smoke drugs than to inject them, but their needs, fears and aspirations were pretty similar to our familiar clients. A study by Anna Edmondson of sixty new heroin clients at two clinics with a mean age of 28 highlighted the three main reasons these addicts might seek help:

1   They hope for an *improvement* in significant, close *relationships* (such as reconciliation with a spouse who has rejected them or regaining custody of children who have been taken into care).
2   They hope to gain or maintain regular *employment*.
3   They hope to receive psychological support for recent traumatic life events (such 'rock bottom' crises included the drug-related death of a spouse).

Aspirations 1 and 2 are pretty common human priorities for anyone aged 28, and accounted for over 90 per cent of the clinic contacts. In Narcotics Anonymous, 'rock bottom' or 'turning point' life crises are often identified, but these differ widely from one individual to another. Unexpected, sudden homelessness, psychiatric illness or domestic violence are examples I have observed of negative life events associated with a fresh 'contemplation' of change, but, sometimes, more positive life events like marriage, pregnancy or even a religious revelation seem to be a turning point which an individual addict says 'changed my heart' (Pullinger 1989).

Occasionally, very different types of need are the antecedent to someone dependent on heroin seeking treatment (such as a medical emergency requiring hospitalization or an impending criminal trial), but non-medical and non-forensic needs like a family, a job and a home are *much* more likely to figure in the wishes of clients wanting to kick the habit in the community.

It is also important to recognize that not everyone contacting a drug service is ready to contemplate a change in their drug intake. One interesting drop-in service in an urban community centre (Kaleidoscope) not only provides an anonymous needle exchange to hundreds of heroin users *to reduce the harm* from unsterile injecting habits, but its friendly volunteers serve cheap meals to users who may just have dropped in for reasons like boredom or loneliness. Even this author, when driving through that neighbourhood, often calls in to experience these colleagues' 'unconditional positive regard' and a late night coffee! An experience of staff who have a good preparation and positive attitudes towards people with dependence can make a big difference to the user's prospects. Not only do *constructive offers of help from an empathic professional* at initial contact greatly increase the probability that addicts subsequently will enter treatment (Raistrick 1989), the *motivational benefits* of making contact with such an empathic professional predict better patient outcomes following that treatment (Miller 1998). These psychological approaches can work at both individual and group levels, in helping users enter and complete treatment effectively (Fiorentine and Anglin 1997).

If one man exemplifies effective engagement with people dependent on drink and drugs, it is Max Glatt, who has been treating addiction continuously since his pioneering work using group therapy in 1952. When the *Observer* newspaper interviewed his former patients for a tribute, these people summed up his approach as follows:

> *He listens, he never patronises, he never judges, he seems genuinely interested in the lives and experiences of those he treats.*
>
> (*Observer*, 17 October 1999)

What blocks engagement with help in many heroin users? At least four problems have been characterized:

1   Dependent users will have many brief periods when they are drug-free (e.g. when their regular dealer has been arrested), but these become associated with a 'pathological preoccupation' (Gorski 1989) in which an addict focuses on the negative aspects of being without drugs and comes to believe that sobriety is awful or terrible. Meanwhile, their intense but unsatisfied *desire* for heroin can even precipitate or accelerate the appearance of *physical withdrawal symptoms* and then, since they are well rehearsed in using heroin to avoid these withdrawal symptoms, they narrow their focus to just how good it would feel to score some more of the drug.

2   A depressed mood is quite common among people dependent on either heroin or cocaine. Poorly recognized but overwhelming feelings of worthlessness, weariness, guilt and shame can interfere with taking *action* such as attending an appointment for treatment.

3   An *external* locus of control (thinking 'my life is like a paper boat in a rainstorm swept along down the gutter by the torrent') and a low self-efficacy (thinking 'nothing I do makes any difference, anyway') undermine efforts to change.

4   Living with a *partner* who is habitually using heroin and who is not
     contemplating any change not only means that changing your own
     habits could threaten that close relationship, but home life will
     present frequent cues for renewed use when you have initially become
     drug free.

Our drug clinic in Cambridge, like most outpatient units, has a high proportion of clients whose dependency on heroin is not reduced over time, even
after repeated efforts to engage in detoxification. An oral substitute for
heroin injecting, methadone, is available on a tapering prescription to
prevent withdrawal symptoms while the users come off street drugs, but
still a number of clients drop in and out of treatment with only brief,
inneffective participation. We now recognize that many of these in-and-
out users are clinically *depressed* and they *attribute* control over their
habit to external factors in a state of resigned helplessness. For about a
quarter of our outpatients, counselling using cognitive and behavioural
skills reduces their depressive symptoms within weeks and their engagement
in detoxification is significantly more successful. After 4 months treatment,
they show much lower measures of dependency than comparable addicts
who had only non-directive counselling without a cognitive-behavioural
framework (Dzialdowski *et al.* 1998). The mutual aid group Narcotics
Anonymous does not usually consider itself a *treatment* agency, but work
by Christo and Sutton (1994) has demonstrated that belonging to 'NA' is
associated with a gradual improvement in self-esteem, with a growing belief
that the recovering addict can take responsibility for their own life. It
follows that many health-oriented treatment services encourage their new
clients also to join this sort of mutual aid group.

In Chapters 16 and 17, a range of treatment and rehabilitation options are
described. One of the key lessons grasped by the new UK anti-drugs strategy
is that being able to offer a variety of treatment *choices* is a very good thing
for the users. Users seeking help can benefit from the staff's skills in negotiat-
ing *individual* treatment goals (Caan 1996). These goals should be reason-
able (in terms of what can be achieved in the short term or in the longer
term) and realistic (in terms of the care facilities available and the personal
strengths and weaknesses brought by the user). The *3 Ws* of patient-centred
care are:

●   What does the user *Want*?
●   What are you *Willing* to provide?
●   What would contribute to the user's *Welfare*?

Acting in the client's welfare

The ethical aspect of engaging with a person dependent on opiates should
not be ignored. Users initially disclose their drug habit for all sorts of

different reasons. In approaching some areas of private medical practice, an addict may wish to supplement his costly and risky supply of street heroin with a regular prescription for pharmaceutically pure opiates for a relatively modest fee. If this continues for a long time, *it is most unlikely to be in the user's interest* in terms of reducing their dependence on the drug or on the doctor. Other ethically dubious areas are interventions which effectively imprison or humiliate the addict to gain control over them. Cults or covertly punitive family members have been known to impose extreme restraint on users. 'Cold turkey' (abrupt withdrawal of opiates) is *not* life threatening, but being stuck in soiled clothes while shaking, sweating, aching and vomiting is not a 'treatment' option many addicts would *choose*. The most dramatic cold turkey can sometimes be precipitated by injecting an opiate *antagonist* like naloxone. Some US services would only prescribe any medication to users for detoxification if they had first submitted to a naloxone 'challenge' and reacted strongly. Hamid Ghodse, scientific advisor to the UN's International Narcotics Control Board, has developed a much more *humane* type of drug 'challenge' to detect a physical dependence on opiates. The topical application of eyedrops containing naloxone to the eye has no effect on the pupil diameter unless there is a long and regular history of opiate use. In people with marked physical dependence, applying such eyedrops to one eye causes a striking but temporary dilation of the pupil on that side, without any distressing symptoms for the client being assessed.

The key question to ask about ethics and addiction is whether the interaction of professional and user is based on the explicit foundation of a therapeutic relationship. Readers who want food for thought might want to look up a case study by George (1998) who describes the engagement of a chaotic 16-year-old heroin injector by a voluntary sector social worker, using clear boundaries and sanctions within a caring relationship. In general, clear goals and mutual honesty provide a good start for relating to drug users. To avoid abusive or exploitative actions, a behaviour 'contract' agreed by both parties can be useful.

A historical perspective on getting into and out of treatment for drug dependence

In the early 1990s, an Acting Chief Commissioner of Police took me aside one evening at a Probation Service public meeting about helping young black people with drug dependence, to whisper in my ear:

Of course, none of them ever really get off it, do they?

He was appallingly wrong, but I can see how his misjudgement might have arisen in the 1970s environment when he would have been a young impressionable officer.

The bad old days

In discussing 'the use of one particular treatment method as if it were applicable to all cases', one London psychiatrist bewailed the return to drug use of about 90 per cent of clients leaving hospital-based units and warned that:

> It is important to emphasise that where the expectation of failure may be high that this can have an important bearing on the outcome of treatment.
>
> (Willis 1969: 83)

In the 1980s, even highly experienced psychiatrists, attempting a gradual detoxification with methadone reduction as the key regime for heroin dependence, found only about 20 per cent were drug-free 6 months after treatment (Menon *et al.* 1986). Single-minded methods of cocaine detoxification were finding a similar one-fifth of patients were drug-free after 6 months (Kang *et al.* 1991). By 1988, when Snyder and Lader were reviewing services for heroin users, there was a growing realization that 'detoxification alone is not an effective treatment'.

During this period, the rapid spread of HIV drew medical attention to the *global* health and family needs of drug users, and family doctors with a *long-term* involvement with their drug-using patients (like Robertson 1989) played a vital role in developing pragmatic and flexible, multi-professional approaches to drug problems. The turning point in the USA came when academic clinicians like Kleber (1989) highlighted the potential contribution of many *diverse* interventions to recovery, with a perspective taking in the whole *life history* of the addict. Thus, many people whose drug use began early in life needed basic *habilitation* in terms of developing usable interpersonal or occupational skills. The value of *family* involvement and developing a structured, rewarding *lifestyle* were recognized and the need to create the right physical *environment* for addicts with polydrug problems or severe psychological disorders was appreciated.

Getting better

In 1991, the popular perception was that growing use of crack cocaine would be the last straw for hopelessly ineffective treatment agencies, because of a myth that it was 'instantly addictive' and led inexorably to the 'disintegration' of users' lives. But two reports that year convinced me that even the dreadful spread of crack could be addressed effectively, when the broader needs of users were considered and interventions were subjected to scientific evaluation. In West Sussex, a creative collaboration between the Probation Service and the Health Service was enabling even criminals with chaotic crack problems to acquire a more internal locus of control and as a result to win some independence from drugs. In London, the CIBA Foundation brought together many stakeholders to consider 'Cocaine: Scientific and Social Dimensions'. Applying the results from basic science studies of cocaine, clinicians were now able to demonstrate some *effective* approaches to both the pharmacotherapy and the psychotherapy of dependence (Bock 1991).

Now that the NTORS study (1997) has followed up its treatment cohort for 6 months, it is clear that both cocaine and heroin problems can be substantially reduced in a variety of treatment settings; for example, methadone reduction services were now effective in getting 38 per cent of heroin users off. The most obvious change in many types of UK service over the last few years has been the growth of inter-professional *teamwork* to address the varied needs of clients, combined with a multi-professional, inter-agency co-operation in developing the quality of practice (Baxter *et al.* 1997). A similar responsiveness to the diversity of users' needs and new effectiveness in psychosocial interventions is being reported in the USA (e.g. in services for Veterans: Stine and Kosten 1997).

Cocaine

The syndrome of dependence on the stimulant drug cocaine has much in common with dependence on the narcotic heroin, above. Over the years, a pattern of compulsive use, tolerance to the effects even of large doses and a growing preoccupation with drug-seeking emerge. There are withdrawal symptoms, but these are different from those seen after opiate withdrawal: typically, sharply alternating periods of exhausted sleep or restlessness and sudden mood swings, which can include intensely suicidal feelings and intense cravings for resumed cocaine use. Because depression and agitation are such prominent features of the first 2 weeks' abstinence from cocaine, the antidepressant drug desipramine is sometimes used during detoxification (Strang *et al.* 1993).

Racing on one roller skate

Stable long-term use of stimulants is extremely difficult to achieve, especially when high doses are administered by injecting cocaine hydrochloride or smoking rocks of the cocaine freebase. Some *opiate* users do achieve stable patterns of use; for example, 40 per cent of the heroin addicts entering the fifteen methadone maintenance programmes in New York are still continuing on a modest dose of oral opioid with little or no use of other drugs, 3 years after beginning maintenance.

- Cocaine dependence usually involves runs of heavy use alternating with exhausted periods of recuperation and the use of other substances (like alcohol, benzodiazepines or cannabis) to relax after the tense, excitable, angry or suspicious states of mind caused by high doses of the stimulant.
- Polydrug use (especially combined bingeing with cocaine and alcohol) is almost inevitable as stimulant use *escalates*. This contrasts with some heroin users who just want their preferred drug and turn to other substances like benzodiazepines, when the heroin supply is interrupted to reduce heroin *withdrawal* symptoms.
- Cocaine addicts tend to use as much of the drug as they have scored, leaving none for the next day. Heroin users often learn to leave a portion of their drug overnight to avoid mild withdrawal symptoms when they wake up the next day.

Different cellular mechanisms

Opiate drugs act directly on nerve cell activity when they bind to enkephalin receptors, mimicking the action of synaptic enkephalins. It is possible to *satisfy* a user's appetite for heroin, when enough brain receptors are occupied by this active drug. Taking a small dose of stimulants like cocaine or nicotine increases the appetite for more drug (Concar 1998). A dependent cocaine user is *insatiable*, and stops using for other reasons than satiety (like running out of money or intrusive, unpleasant, anxious sensations or even seizures kindled by frequently repeated doses). Substituting a different stimulant like prescribed methamphetamine or methylphenidate does not lead to stability – it just primes the addict to seek more. Cocaine has *no direct effect* on the electrical signals in nerve cells (Uchimura and North, 1990), but if dopamine is being released then this transmitter accumulates to concentrations far above normal physiological levels as cocaine prevents the brain's dopamine re-uptake systems from retrieving the transmitter after its release. All the effects of dopamine are dramatically amplified, whether the post-synaptic dopamine receptors have an excitatory or an inhibitory action. Brain PET scans in regular cocaine users have shown that people *only* report a drug-induced 'high' when most of the re-uptake sites have been put out of action (Volkow *et al.* 1997) and, so, to sustain the flood of extracellular dopamine in their brains with a short-acting drug like cocaine, frequently repeated doses are required. Unfortunately, chronically abnormal dopamine levels lead to a loss of dopamine *receptors* on precisely those nerve cells which are the substrate of cocaine's psychological effects

(Volkow *et al.* 1990), so the euphoric effect of cocaine achieved in a naive user can never be recaptured in an addict, no matter how much of the drug is used.

Cocaine and learning mechanisms

Rats learning to work for a small injection of cocaine may come to devote so much effort to self-administering more and more of this drug that they will ignore weariness, hunger and even drug-induced fits to work themselves to death for cocaine. Structures in the basal ganglia of the brain (especially the nucleus accumbens in rats) play a part in the rewarding effects of systemic cocaine injections, but direct intracranial administration into the medial prefrontal cortex also provides a strong reinforcement for learning (Goeders and Smith 1983). Goldman-Rakic (1998) has demonstrated 'memory fields' corresponding to behavioural working memories, in the prefrontal cortex of the monkey. Dopamine facilitates sensory signals in stimulating these memory circuits. Cocaine in people with dependence not only acts as a primary reinforcer, but turns many stimuli associated with drug use into secondary reinforcers, which, in their turn, come to cue a growing catalogue of drug-related behaviours (Caan 1993). These *cueing* effects happen to some extent with a growing alcohol or heroin dependence, but they are most prominent in developing dependence on cocaine; for example, a meta-analysis of research on cue responses in different types of dependence confirmed that the worst cravings were experienced for cocaine (Carter and Tiffany 1999).

Cocaine addicts may find it easier than people dependent on heroin to *stop* taking their drug for a short period, but the risk of a *relapse* to cocaine use is very high. Stasiewicz and Maisto (1993) have proposed a learning model of drug dependence which combined the development of conditioned responses (CRs) to the positive aspects of the drug with the avoidance of negative emotional states (such as fear). Supposing an addict has not taken any drugs for a while . . .

> following extinction of drug CRs to drug-related stimuli, the abstinent individual will remain vulnerable to relapse, because the conditioned emotional responses to aversive stimuli will not have been extinguished.
> (Stasiewicz and Maisto 1993: 347)

Negative feelings have been reported as a common antecedent in many studies of relapse to drug use. What suggests that the two-factor model of Stasiewicz and Maisto may be especially relevant to cocaine addiction is the observation of Kasarabada *et al.* (1998) that the men whose initial cocaine use progresses most rapidly to severe dependence were those men who had the most problems *in situations involving unpleasant emotions.*

One lesson that services are now learning, in trying to help people with cocaine dependence, is to respond to sudden downswings in mood 'simply and quickly' (Mapp 1997). As the Crack Outreach Team in Birmingham observed, 'if you're all wired up after a dose of crack there's no point offering you an appointment at a centre in two weeks time.'

A test

You are taking two members of NA out to the cinema, after one has just lost her job (feeling anger and resentment). Her partner picks an action film about gangsters, who turn out to be drug traffickers. He had a history of heroin use but she used cocaine (along with many other substances). She has been 'clean' every day for 5 years (i.e. drug-free) when she witnesses a cinema screen filled for several minutes with scenes of cocaine powder being converted to crack in a dealer's head-quarters. Becoming increasingly distressed, shaking and trying to hide from the stimuli, the former user eventually retreats to the toilets with strong drug cravings. Soon after this episode, she has a brief lapse to drug use. How would you address the burden of her 'unforgettable' memories during the next few days?

One of the research areas currently showing promise is that of relapse prevention, although it may sometimes be difficult in a clinic setting to anticipate cues that might arise in the community (like a cinema). Meanwhile, one of the techniques which NA recommends to its newly drug-free members, to build up their resilience in the face of pressures, is to avoid the 'HALT' feelings: Hungry, Angry, Lonely or Tired. Mutual aid can sometimes help with all four vulnerable HALT states.

Understanding personal stories

Recovery from dependence on drugs is a very personal learning experience. Using 'narratives of recovery' can, however, help professionals to see patterns in the routes to well-being people that people explore; for example, a recent Finnish study found five main 'story types' among twenty-two men and twenty-nine women recovering from a variety of substances or gambling (Hanninen and Koski-Jannes 1999). From the point of view of the drug users, understanding such narratives means that 'all our loss and pain will not be in vain' (Efthimiou-Mordaunt 1998).

Special people with special needs

The Local Government Association reported in 1998 on the need for specialist support services for young drug users in children's homes and for those about to leave care.

There is a growing awareness that small communities, which rely on one employment like fishing (Munro 1999), can face sudden increases in drug problems when these fisheries collapse, as is happening all along the North Sea coast in 2001.

The need for improved services for adults in prison also raises the question of how we should establish therapeutic relationships in custodial settings, and how to prepare for life transitions as people return to the community.

The most controversial group with special care needs are dual-diagnosis patients with both a mental illness and drug dependence. In their 1998 report *The Unlearned Lesson*, Alcohol Concern and the Standing Conference on Drug Abuse combined to describe the possible role of drugs (e.g. combined cocaine and alcohol use) in homicides from cases in recent psychiatric inquiries. Aggression is generally more common among those psychiatric patients who also have a problem with substance use (Scott *et al.* 1998) and there is sometimes a fear among services that such cases are too hot to handle. However, some services like the day hospitals studied by Galanter *et al.* (1993) or Caan *et al.* (1996) seem to engage well with dual-diagnosis clients.

Another group who were largely left out of the recent UK drugs strategy are pregnant women with drug dependence. A high proportion of these pregnant users have endured sexual assaults (72 per cent) or physical assaults (67 per cent), and, if a resulting post-traumatic stress disorder is present, this renders it difficult to engage in treatment (Thompson and Kingree 1998). The *nature* of their dependence needs to be seen within the context *both* of their biology (pregnancy) *and* their life story (abuse).

References

Abbasi, K. (1998) 'Deaths from heroin overdose are preventable', *Brit. Med. J.*, **316**, 331.

Adams, K. (1998) 'A tale of four cities', *Health Service J.*, 5 February, 26–7.

Allen, D. F. and Jekel, J. F. (1991) *Crack: The Broken Promise*. London: Macmillan.

Argyle, M. (1992) *The Social Psychology of Everyday Life*. London: Routledge.

ANSA (1997) *Substance Use: Guidance on Good Clinical Practice for Specialist Nurses*. London: Association of Nurses in Substance Abuse.

Baxter, B., Green, S., Williams, O. and Caan, W. (1997) 'Injecting some audit into substance misuse services', *Psychiat. Bull.*, **21**, 360–3.

Bean, P. and Winterburn, D. (1997) 'Persistent drug-misusing offenders', *ARVAC Bull.*, **67**, 15.

Benton, D. (1988) 'The role of opiate mechanisms in social relationships', *The Psychopharmacology of Addiction*, pp. 115–40. Oxford: Oxford Medical Publications.

Blake, R. and Stephens, E. (1987) *Compulsion*. London: Thames TV and Boxtree.

BMJ (1998) 'One in 15 doctors in UK may suffer some form of dependence', *Brit. Med. J.*, **316**, 328.

Bock, G. (1991) 'Cocaine: Scientific and social dimensions', *Ciba Foundation Bulletin*, **30**, 13–14.

Bradley, A. and Mounteney, J. (1998) 'Tsars in their eyes', *Community Care*, 5 February, 22–3.

Caan, W. (1993) 'Learning mechanisms in cocaine dependence', *The Society for the Study of Addiction Annual Symposium Abstracts*, p. 10.

Caan, W. (1994) 'Community psychiatric nursing: Drug-use and HIV-risk characteristics', *J. Psychiat. Mental Health Nurs.*, **1**, 49.

Caan, W. (1996) 'Clinical effectiveness and patient centred care', *Putting Evidence Based Medicine into Practice*. London: IBC.

Caan, W., Rutherford, J., Carson, J., Holloway, F. and Scott, A. M. (1996) 'Auditing psychiatric day hospitals', *J. Mental Health*, **5**, 173–82.

Carter, B. L. and Tiffany, S. F. (1999) 'Meta-analysis of cue-reactivity in addiction research', *Addiction*, 94, 327–40.

Chemicon (1999) 'Antibodies to cocaine- and amphetamine-related transcript (CART)', *Communications Update*, **9**(6), 1–2.

Childress, A. R., Ehrman, R., McLellan, A. T., MacRae, J., Natale, M. and O'Brien, C. P. (1994) 'Can induced moods trigger drug-related responses in opiate abuse patients?', *J. Substance Abuse Treat.*, **11**, 17–23.

Christo, G. and Sutton, S. (1994) 'Anxiety and self-esteem as a function of abstinence time among recovering addicts attending Narcotics Anonymous', *Brit. J. Clin. Psychol.*, **33**, 198–200.

Clark, G. (1998) 'Trick or treatment', *Nurs. Times*, **94**(6), 12–13.

Concar, D. (1998) 'Lust', *New Scientist*, **157**(2127), 40–3.

Coombs, R. H. (1996) 'Addicted health professionals', *J. Substance Misuse for Nurs. Health Social Care*, **1**, 187–94.

Cottler, L. B. and Compton, W. M. (1993) 'Advantages of the CIDI family of instruments in epidemiological research of substance use disorders', *Int. J. Meth. in Psychiatr. Res.*, **3**, 109–19.

Damasio, A. R. (1997) 'Towards a neuropathology of mood', *Nature*, **386**, 769–70.

Davidson, E. S., Finch, J. F. and Schenk, S. (1993) 'Variability in subjective responses to cocaine: Initial experiences of college students', *Addict. Behav.*, **18**, 445–53.

Davies, J. (1997) 'Addiction myth out of control', Interview, *Times Higher Education Supplement*, 13 June, 7.

Dzialdowski, A., London, M. and Tilbury, J. (1998) 'A controlled comparison of cognitive behavioural and traditional counselling in a methadone tapering programme', *Clin. Psychol. Psychother.*, **5**, 47–53.

Efthimiou-Mordaunt, A. (1998) 'All our loss and pain will not be in vain', *The User's Voice*, **1**(3), 1.

Faithfull, M. (1994) 'Quote ... unquote', *Int. J. Drug Policy*, **5**, 203.

Farmer, R. (1997) 'Substance misuse: Aspects of treatment and service provision', *Adv. Psychiat. Treat.*, **3**, 174–81.

Fiorentine, R. and Anglin, M. D. (1997) 'Does increasing the opportunity for counselling increase the effectiveness of outpatient drug treatment?', *Amer. J. Drug Alcohol Abuse*, **23**, 369–82.

Frankel, S. (1998) 'The fashion of destruction', *Guardian*, 7 February, 5.

Fugelstad, A., Annell, A., Rajs, J. and Agren, G. (1997) 'Mortality and causes and manner of death among drug addicts in Stockholm during the period 1981–1992', *Acta Psychiatrica Scandinavica*, **96**, 169–75.

Galanter, M., Egelko, S., De Leon, G. and Rohrs, C. (1993) 'A general hospital day program combining peer-led and professional treatment of cocaine abusers', *Hospital Community Psychiat.*, **44**, 644–9.

George, M. (1998) 'The risk factor. Too much, too young', *Community Care*, 2 April, 30–1.

Gerada, C. and Ashworth M. (1997) 'Addiction and dependence – I: Illicit drugs', *Brit. Med. J.*, **315**, 297–300.

Ghodse, H. (1989) *Drugs and Addictive Behaviour*. Oxford: Blackwell.

Goeders, N.E. and Smith, J. E. (1983) 'Cortical dopaminergic involvement in cocaine reinforcement', *Science*, **221**, 773–5.

Goldman-Rakic, P. (1998) 'Regional and cellular fractionation of working memory', *Brit. Neurosci. Assoc. Newsletter*, **31**, 4–5.

Gorski, T. T. (1989) *The Relapse/Recovery Grid*. Center City, MN: Hazelden.

Government Statistical Service (1998) *Drug Misuse Statistics for Six Months Ending September 1996*. London: Department of Health.

Guirguis, J. and Vostanis, P. (1998) 'Keeping score', *Community Care*, 12 February, 30–31.

Haarhoff, G. and London, M. (1995) 'A comparative study of injecting opiate and amphetamine users in a rural area', *Addict. Res.*, **3**, 33–8.

Handelsman, L., Branchey, M. H., Buydens-Branchey, L., Gribomont, B., Holloway, K. and Silverman, J. (1993) 'Morbidity risk for alcoholism and drug abuse in relatives of cocaine addicts', *Amer. J. Drug Alcohol Abuse*, **19**, 347–57.

Hanninen, V. and Koski-Jannes, A. (1999) 'Narratives of recovery from addictive behaviours', *Addiction*, **94**, 1837–48.

Hellawell, K. (1998) 'Report of the UK Anti-Drugs Coordinator', *Tackling Drugs to Build a Better Britain*, pp. 6–36. London: Her Majesty's Stationery Office.

Hirst, J. (1999) 'Losing the using battle?', *Community Care*, 23 September, 22–3.

Hubner, C. B. and Koob, G. F. (1990) 'The ventral pallidum plays a role in mediating cocaine and heroin self-administration in the rat', *Brain Res.*, **508**, 20–9.

Jellinek, E. M. (1942) *Alcohol Addiction and Chronic Alcoholism*. New Haven, CT: Yale University Press.

Jurith, E. H. (1998) 'It's no quick fix', *Guardian Society*, 13 May, 6–7.

Kang, S-Y., Kleinman, P. H., Woody, G. E., Millman, R. B., Todd, T. C., Kemp, J. and Lipton, D. S. (1991) 'Outcomes for cocaine abusers after once-a-week psychosocial therapy', *Amer. J. Psychiat.*, **148**, 630–5.

Kasarabada, N. D., Anglin, M.D., Khalsa-Denison, E. and Paredes, A. (1998) 'Variations in psychosocial functioning associated with patterns of progression in cocaine-dependent men', *Addict. Behav.*, **23**, 179–89.

Kleber, H. D. (1989) 'Treatment of drug dependence: What works', *Int. Rev. Psychiat.*, **1**, 81–100.

Kroll, B. (1997) 'The sins of the mothers', *Guardian Society*, 20 August, 2–3.

McCrone, J. (1997) 'Wild minds', *New Scientist*, 13 December, 26–30.

McLellan, A. T., Luborsky, L. Woody G. E. and O'Brien, C. P. (1980) 'An impoved diagnostic evaluation instrument for substance abuse patients', *J. Nervous Mental Dis.*, **168**, 26–33.

Mapp, S. (1997) 'Wired up', *Community Care*, 4 December, 27.

Mapp, S. (1998) 'Learning better habits', *Community Care*, 5 March, 22.

Markou, A., Weiss, F., Gold, L. H., Caine S. B., Schulteis G. and Koob, G. F. (1993) 'Animal models of drug craving', *Psychopharmacology*, **112**, 163–82.

Menon, P., Evans, R. and Madden, J. S. (1986) 'Methadone withdrawal regime for heroin misusers: Short-term outcome and effect of parental involvement', *Brit. J. Addict.*, **81**, 123–6.

Miller, W. R. (1998) 'Why do people change addictive behaviour?', *Addiction*, **93**, 163–172.

MRC (1994) *The Basis of Drug Dependence*. London: Medical Research Council.

Munro, R. (1999) 'Cod and cold turkey', *Nurs. Times*, **95**(24), 14–15.

Murtra, P., Sheasby, A. M., Hunt, S. P. and de Felipe, C. (2000) 'Rewarding effects of opiates are absent in mice lacking the receptor for substance P', *Nature*, **405**, 180–3.

NTORS (1996) *The National Treatment Outcome Research Study*. London: Department of Health.

NTORS (1997) *The National Treatment Outcome Research Study*, 2nd Bulletin. London: Department of Health.

OPCS (1994) *OPCS Surveys of Psychiatric Morbidity in Great Britain*, Bulletin 1. London: OPCS Social Survey Division.

Petry, N. M. and Bickel, W. K. (1998) 'Polydrug abuse in heroin addicts: A behavioral economic analysis', *Addiction*, **93**, 321–35.

Prince of Wales, HRH (1998) 'A manifesto of hope', *The Big Issue*, 16 February, 11.

Pullinger, J. (1989) 'Ah Kam', *Crack in the Wall*, pp. 193–4. London: Hodder & Stoughton.

Raistrick, D. (1989) 'Making treatment decisions', *Int. Rev. Psychiat.*, **1**, 173–80.

Robertson, J. R. (1989) 'Treatment of drug misuse in the general practice setting', *Brit. J. Addiction*, **84**, 377–80.

Scott, H., Johnson, S., Menezes, P., Thornicroft, G., Marshall, J., Bindman, J., Bebbington, P. and Kuipers, E. (1998) 'Substance misuse and risk of aggression and offending among the severely mentally ill', *Brit. J. Psychiat.*, **172**, 345–50.

Snyder, S. H. and Lader, M. H. (eds, 1988) *Heroin. The Street Narcotic*. London: Burke.

Social Services Inspectorate (1997) *Substance Misuse and Young People: The Social Services Response*. London: Department of Health.

Stanley, J. and Godsmark, A. (1997) *The Health of the Young Nation*. Norwich: Norfolk Health.

Stasiewicz, P. R. and Maisto, S. A. (1993) Two-factor avoidance theory: The role of negative affect in the maintenance of substance use and substance use disorder. *Behav. Ther.*, **24**, 337–56.

Stine, S. M. and Kosten, T. R. (eds, 1997) *New Treatments for Opiate Dependence*. London: Guilford Press.

Strang, J., Johns, A. and Caan, W. (1993) 'Cocaine in the UK – 1991', *Brit. J. Psychiat.*, **162**, 1–13.

Thompson, M. P. and Kingree, J. B. (1998) 'The frequency and impact of violent trauma among pregnant substance abusers', *Addict. Behav.*, **23**, 257–62.

Uchimura, N. and North, R. A. (1990) 'Actions of cocaine on rat nucleus accumbens neurones in vitro', *Brit. J. Pharmacol.*, **99**, 736–40.

Volkow, N. D., Fowler, J. S., Wolf, A. P., Schyler, D., Shiue, C-Y., Alpert, R., Dewey, S. L., Logan, J., Bendriem, B., Christman, D., Hitzemann, R. and Henn, F. (1990) 'Effects of chronic cocaine abuse on postsynaptic dopamine receptors', *Amer. J. Psychiat.*, **147**, 719–24.

Volkow, N. D., Wang, G-J., Fischman, M. W., Foltin, R. W., Fowler, J. S., Abumrad, N. N., Viktun, S., Logan, J., Gatley, S. J., Pappas, N., Hitzemann, R. and Shea, C. E. (1997) 'Relationship between subjective effects of cocaine and dopamine transporter occupancy', *Nature*, **386**, 827–30.

Williams, O., Cassidy, J. and Caan, W. (1996) 'Banged up, shooting up, messed up', *Health Service J.*, 27 June 1996, 26–7.

Willis, J. H. (1969) *Drug Dependence*. London: Faber and Faber.

Willis, J. H. P. (1977) 'Drug addiction', *The Encyclopaedia of Ignorance*, pp. 369–72. Oxford: Pergamon Press.

Wilson, J. and Campbell, D. (1998) 'There's no youth club, the nearest town is 20 miles away, and you can't go to the pub because your parents are there', *Guardian* G2, 11 March, 2–3.

Chapter 15

Benzodiazepine abuse

Heather Ashton

Key points

- *Benzodiazepine abuse is a growing problem and carries serious risks to health and society.*
- *Benzodiazepines are commonly used by polydrug abusers, alcoholics and sometimes as primary recreational drugs.*
- *People who abuse benzodiazepines often take very large doses orally, by injection or by snorting.*
- *Benzodiazepine use leads to dependence and a withdrawal syndrome which may include convulsions and psychosis.*
- *Further research is needed on the optimal short-term and long-term management of benzodiazepine abuse.*
- *The primary source of illicit benzodiazepines is from doctors' prescriptions.*

Benzodiazepines are consumed by two main populations with different characteristics: (1) low-dose prescribed benzodiazepine users and (2) high-dose, non-prescribed benzodiazepine abusers. This chapter is concerned with the second group and also with a smaller intermediate group of high-dose prescribed users, some of whom become involved in illicit use. While the prevalence of prescribed dose benzodiazepine use is declining, that of illicit use has been rising steeply since the 1980s and now presents a major health problem.

Who abuses benzodiazepines?

Benzodiazepines are taken for recreational purposes by increasing numbers of drug abusers (Drug and Therapeutics Bulletin 1997). The true prevalence is not known but benzodiazepines commonly form part of a polysubstance abuse pattern; for example, at the Liverpool Drug Dependency Unit, 44 per cent of a random sample of 100 injecting drug users entering treatment were also using benzodiazepines (Shaw *et al*. 1994). Benzodiazepines have been taken by opiate, amphetamine and cocaine users worldwide for about 20 years and are now creeping into the teenage 'rave' scene amongst users of MDMA (Ecstasy) and LSD (Strang *et al*. 1993). Various intoxicating drug–benzodiazepine combinations such as 'Tem-Tems' (buprenorphine and temazepam) and temazepam and lager are popular, particularly in the north of England and Scotland. Indeed, benzodiazepines are the single most misused category of drug in Scotland (Robertson and Ronald 1992). A contemporary youth craze in Glasgow is to ride the buses all day 'high' and 'wobbling' on oral temazepam, fortified by high-strength lager and/or cannabis (Parrott 1995).

To begin with, as with alcohol, smoking and cannabis, some of these youths become involved in recreational benzodiazepine use at 13 and 14 years of age (Wilkinson 1997; Pedersen and Lavik 1991). The age range of benzodiazepine abusers surveyed in various British addiction clinics was 19–31 and the male-to-female ratio was between 2.8:1 and 2.1:1 (Strang *et al*. 1994; Ruben and Morrison 1992).

Second, benzodiazepine abuse is also common in alcoholics. Around 30–50 per cent of alcoholics attending for detoxification also use non-prescribed benzodiazepines (Borg *et al*. 1993).

Third, some people (again an unknown number) use benzodiazepines as their primary recreational drug, typically bingeing intermittently on high doses (Strang *et al*. 1993).

It has been claimed that benzodiazepine abuse is 'of little or no consequence' in the vast population of prescribed benzodiazepine users (Woods *et al*. 1988). However, a proportion of prescribed users do escalate dosage, take the drugs for hedonic effects and enter into the illicit drug scene (see Case 15.1). It is important to remember that substance misuse can occur across a wide spectrum of ages and not to allow prejudice to hinder older people's diagnosis and treatment (Greenwood 2000).

Case 15.1 'The one that got a way': escalation from prescribed to street use of temazepam

Peter is the youngest of nine siblings. His mother is alcoholic; his father is unknown. He was taken into care at the age of 2 and brought up in a series of children's homes. He was said to be a quiet and sensitive child who was always terrified of violence. When he was about 13, following an incident with one of the teachers at a boarding school, Peter 'discovered' he was gay.

In his 20s, Peter was involved, as a passenger on a motor cycle, in a road traffic accident. He sustained serious injuries including fractures of his arm and collar bone and compound fractures in his leg. He spent several months in an orthopaedic ward and suffered complications including infection, non-healing of the leg fractures, bone-grafting, insertion of pins and plates, etc.

In hospital, he experienced 'dreadful panic attacks and fears'. He received no psychological treatment or counselling of any kind but was prescribed temazepam and dihydrocodeine. At first, the dosage of temazepam was 20 mg nocte for sleep, but was increased to 60 mg in hospital. After discharge, he continued to receive temazepam from his general practitioner because of panics and insomnia and the dosage was increased over a period of years until he was taking 80 mg temazepam each night and 40–80 mg during the day. He felt he had to take the temazepam, otherwise he got panicky and frightened, suffered pains in the stomach and was unable to sleep. After taking temazepam, he stated that he felt 'all nice and calm for a time, but then the panics and fear return'.

At the age of 30, Peter was removed from his GP's practice when it was discovered that he had altered a prescription for temazepam. He was taken on by another GP but later attended a series of GPs, obtaining ever larger prescriptions of temazepam and dihydrocodeine from each, often presenting with stories that his last prescription had been lost or stolen. He also attempted to obtain supplies of temazepam from the hospital pharmacy, the orthopaedic clinic and the casualty department, sometimes wearing a white tunic and name badge, giving the impression he was a nurse.

When he could no longer satisfy his need from prescriptions, Peter took to obtaining temazepam on the street, taking large and irregular doses by mouth. His behaviour became chaotic and he was twice sent to prison for credit card frauds. In prison, he was terrified and made bizarre claims about his health: that he was on renal dialysis, that he was HIV-positive, that he expected to have his leg amputated through disease. He was able to obtain temazepam in prison from other prisoners.

After discharge, and now aged 34, Peter agreed to enter a detoxification centre and apparently made a real effort to stop the temazepam. Initially, he made good progress. The temazepam was replaced with decreasing doses of diazepam and following his admission he continued for some months to attend the centre weekly as an outpatient. Withdrawal symptoms consisted of increasing anxiety, panics and stomach pains. He took no other drugs, as confirmed by weekly urine tests, and very little alcohol. He had never injected any drugs. Unfortunately, when down to only 4 mg diazepam daily, he 'broke his contract' and obtained some temazepam on the street. This resulted in immediate discharge from the detoxification centre and no further medical supervision.

When last heard of, Peter, aged 35, was again buying temazepam illicitly and was involved in a court case for obtaining money on false pretences.

Comments

1   Initially prescribed benzodiazepines, if not carefully supervised, can lead to escalation of dosage and entry into the illicit drug scene in vulnerable individuals.
2   Strict contract methods are not always appropriate for benzodiazepine-dependent patients. If Peter had been given another chance by the drug clinic (having come so far in his treatment), he might have been able to break his dependence.
3   The possibility of a maintenance dose of diazepam, rather than complete withdrawal, could perhaps have been considered in this case.

Why abuse benzodiazepines?

The most common reason given by polydrug abusers for taking benzodiazepines is that they enhance and often prolong the 'high' obtained from other drugs including heroin, other opioids, cocaine and amphetamines. Benzodiazepines are mainly taken along with the primary drug, but sometimes used alone as an alternative or in times of shortage. Second, benzodiazepines alleviate withdrawal effects, including anxiety and insomnia, when supplies of other drugs are limited. Users of stimulants including cocaine, amphetamines and Ecstasy also take benzodiazepines as 'downers' to overcome the effects of their 'uppers' and to combat hangover effects. In alcoholics, benzodiazepines are used partly to alleviate the anxiety associated with chronic alcohol use, but also because the mixture of alcohol and benzodiazepines produces a hedonic effect. Finally, benzodiazepines, when taken alone in high doses and particularly when injected, can themselves provide a 'kick'.

Although benzodiazepines in therapeutic doses have been claimed to have little abuse potential compared with other drugs of abuse (Warburton 1988), their abuse liability may vary along the dose–response curve, becoming greater at doses above the therapeutic range (Strang *et al.* 1993). Diazepam, alprazolam, lorazepam and triazolam have all been shown in clinical laboratory studies to possess abuse liability. Alprazolam 1 mg was comparable with 10 mg d-amphetamine in scores for 'elation' and 'abuse potential' in experienced but non-dependent users. 'High' ratings for oxazepam and

Table 15.1 Some benzodiazepines used recreationally (UK, Europe and USA)

Generic name	Brand name (UK)	Potency (approximate dose equivalent to 10 mg diazepam)
Alprazolam	Xanax	0.5
Bromazepam	Lexotan	5
Chlordiazepoxide	Librium	25
Diazepam	Valium	10
Flunitrazepam	Rohypnol	1
Flurazepam	Dalmane	15–30
Ketazolam[1]	Anxon	15–30
Lorazepam	Ativan	1
Medazepam[1]	Nobrium	10
Nitrazepam	Mogadon	10
Oxazepam	Serenid	20
Prazepam[1]	Centrax	15
Temazepam	Normison, Euhypnos	20
Triazolam[1]	Halcion	0.5
(Zopiclone)[2]	(Zimovane)	(15)
(Zolpidem)[2]	(Stilnoct)	(20)

Notes
1 No longer in British National Formulary.
2 Non-benzodiazepine hypnotics with similar actions to benzodiazepines; may have abuse liability.

chlordiazepoxide were lower than for other benzodiazepines (Zawertailo *et al.* 1995; Griffiths and Wolf 1990). The non-benzodiazepine hypnotic zolpidem produced 'drug liking' scores similar to triazolam (Evans *et al.* 1990); this drug and the similar non-benzodiazepine hypnotic zopiclone may also have abuse potential.

Which benzodiazepines are abused?

Nearly all the available benzodiazepines have been abused (Table 15.1). In general, those which enter the brain rapidly (e.g. diazepam) are preferred to those which are absorbed more slowly (e.g. oxazepam) (Griffiths *et al.* 1984). However, preferred drugs vary between countries and over time depending on their availability and reputation in the illicit drug world. In the UK, temazepam has superseded diazepam, nitrazepam and flurazepam as the most commonly abused benzodiazepine, in line with the increase in temazepam prescriptions and possibly (until recently) because of the availability of easily injectable forms of temazepam from capsules, 'jellies', 'eggs' (Stark *et al.* 1987). In the US, flunitrazepam tablets ('roaches') have become popular, partly because of diversion of supplies across the Mexican border (Roche Pharmaceuticals report 1996). Potent benzodiazepines such as triazolam (no longer available in the UK), alprazolam (widely prescribed in the US) and lorazepam have also achieved popularity among benzodiazepine abusers.

Routes of administration and dosage

Benzodiazepines can be taken by mouth, inhaled as snuff or injected. The commonest practice is oral ingestion but, recently, novel forms of administration have been used. Intranasal 'snorting' of powdered flunitrazepam is described (McNamee 1994) and this method can be used for other benzodiazepines and accompanying drugs such as buprenorphine (Strang 1991). However, the main alternative to the oral route is intravenous injection, reported first for flurazepam (Strang 1984) and now increasingly practised in the UK. Diazepam and other benzodiazepines have been injected but at present temazepam is mainly involved. Strang *et al.* (1994) conducted a questionnaire survey of subjects attending drug clinics in seven British cities. Of 208 subjects returning the questionnaire, 186 had used benzodiazepines and 103 had injected them intravenously. Temazepam was the most commonly used and had been injected from preparations of capsules, tablets and syrup. Other injected benzodiazepines were diazepam, lorazepam, triazolam, nitrazepam and chlordiazepoxide. Attempts to discourage temazepam injections by substituting liquid with gel-filled capsules and by introducing tablets and elixir appear to be unsuccessful since the gel can be warmed to a liquid consistency, the tablets can be dissolved in warm water (Carnwath 1993) and the elixir diluted to provide injectable solutions.

Doses used by recreational benzodiazepine users are usually far in excess of those recommended for therapeutic purposes. Oral and intravenous doses of 100–150 mg temazepam and diazepam are common (Robertson and Ronald 1992) while some youths may take up to fifty tablets of temazepam (500–1,000 mg) for hedonic effects (Parrott 1995). A patient described by Cole and Chiarello (1990) had been taking 10 mg alprazolam several times a day for its euphoriant actions. In twenty-three intravenous temazepam abusers, the mean daily dose injected was 610 mg; the maximum was 3,600 mg (Ruben and Morrison 1992). Farrell and Strang (1988) reported a polydrug user who had been injecting the contents of 30–90 temazepam (20 mg) capsules daily. Subjects become remarkably tolerant to the sedative/hypnotic effects of benzodiazepines and the latter patient had no major alterations in consciousness (see also Case 15.2).

Case 15.2 Inappropriate benzodiazepine prescribing after alcohol detoxification

Despite her name, Joy was an unloved child. Her mother told her: 'The biggest mistake you made in life was being born.' She was physically abused by her stepfather and raped by a gang of youths when she was 16. She ran away from home and worked for some years in hotels. By the age of 21, she had a serious alcohol problem. She was rarely sober, suffered from blackouts and blood tests showed early liver damage.

Eventually, aged 24, Joy entered a clinic for alcohol detoxification, followed by prolonged psychotherapy and counselling. She was successful in stopping drinking, but, because of severe anxiety, insomnia and social phobia, the benzodiazepines initially prescribed during alcohol detoxification were continued by the psychiatrist and the dosage increased. When referred to a tranquillizer withdrawal clinic 2 years later, Joy was taking prescribed doses of 240 mg oxazepam and 240 mg diazepam daily – amounting to forty-eight tablets each day. In spite of this dosage, she was able to ride a bicycle and attend a writing course, but still suffered from anxiety and insomnia and seemed to exist in a permanent daze.

Nevertheless, she was keen to withdraw from benzodiazepines as they were dominating her life. Her course over the next 3 years was stormy. She made repeated suicide gestures with overdoses and with self-injury. She became depressed and the slightest attempts at dosage reduction were followed by panic attacks and threats of suicide. After four planned 2-week admissions to hospital for moderate dosage reduction, she eventually became drug-free. Withdrawal symptoms, which were severe, included depersonalization, tremor, muscle pain and stiffness, palpitations, blurred vision, nightmares and increased aggression. She received the SSRI sertraline during the last 6 months of withdrawal but this was later tapered off and stopped. A psychiatric assessment performed by a different psychiatrist at that time suggested borderline personality disorder.

One year after withdrawal, Joy, now aged 30, is no longer depressed, sleeps well and has lost much of her anxiety although she still suffers from social phobia for which she is receiving cognitive therapy. Neuropsychological testing revealed some frontal lobe dysfunction which was attributed to the heavy alcohol/benzodiazepine use, but Joy is now actively seeking employment.

Comments

1   Prescriptions for benzodiazepines during alcohol detoxification can lead to benzodiazepine dependence.
2   Chronic use of benzodiazepines can lead to a remarkable degree of tolerance to their hypnotic, anxiolytic and ataxic effects.
3   Patients who have taken large doses of benzodiazepines for several years can be withdrawn if sufficiently motivated and carefully handled, and their psychological state may improve.

> 4   Patients dependent on large prescribed doses of benzodiazepines
>   do not necessarily progress to illicit use.
>
> *Discussion Point*
> To treat someone with multiple problems successfully, it can be
> valuable to take a 'long' view (see Chapter 17). How would you
> take account of Joy's early experience of family turmoil, her past
> use of alcohol, her vulnerability to depression and especially her
> previous experience of frightening symptoms during drug withdrawal
> (Moos *et al.* 2002)?

Sources of benzodiazepines

Benzodiazepines are widely available on the street and are cheap (about £1 a
tablet in Newcastle-upon-Tyne at present). A major source is from general
practitioners' prescriptions. Some users attend several practitioners using
false names and temporary patient status; others obtain supplies from
friends or patients (often elderly people) who exaggerate their needs to
their doctors and sell off the excess (Ruben and Morrison 1992). Some
children obtain them from prescriptions for their parents (Pedersen and
Lavik 1991). Benzodiazepines are also obtained by theft from health
centres or retail chemists, and large amounts have been stolen from pharmaceutical warehouses. In 1996, 29,000 temazepam tablets were known to be
stolen from such a warehouse in Newcastle-upon-Tyne and 30,000 diazepam
tablets from a similar source in Cumbria (Newcastle police information).

Health risks and social consequences

Some hazards associated with high-dose benzodiazepine abuse are shown in
Table 15.2. These risks have probably been underestimated. Benzodiazepines are generally believed to be safe in overdose, but deaths following
self-poisoning do occur, even when the drugs are taken alone, and a fatal
outcome from overdose is more likely with flurazepam and temazepam than
with other benzodiazepines (Serfaty and Masterton 1993). Benzodiazepines
also add to the respiratory depression caused by other drugs: the combination of temazepam with injected opioids (e.g. buprenorphine) is said to
cause approximately 100 deaths a year in Glasgow alone (Parrott 1995).

Benzodiazepine use increases the risk of road traffic accidents, especially
when driving under the influence of higher doses (Bandolier 1998).

Mental disturbances caused by benzodiazepines include blackouts and
memory loss, aggression, violence and chaotic behaviour associated with
paranoia. Over a third of intravenous temazepam users surveyed by

Table 15.2 Health and social hazards of benzodiazepine abuse

General	Complications of IV use
Fatalities due to overdose (particularly in combination with opioids)	Thrombophlebitis
	Deep and superficial abscesses
Blackouts and memory loss	Deep vein thrombosis
Paranoia	Pulmonary microembolism
Violence and criminal behaviour	Rhabdomyolysis, tissue necrosis
Risk-taking sexual behaviour	Gangrene, requiring amputation (usually
Long-term cognitive impairment	due to inadvertent intra-arterial
Foetal and neonatal risks if taken in	injection)
pregnancy	Hepatitis B and C
Dependence	HIV infection
Withdrawal seizures	

Ruben and Morrison (1992) used the drug 'to give confidence to engage in criminal activity', and there is a high incidence of illicit benzodiazepine use in remand prisoners (Mason et al. 1997). The loss of judgement and amnesia caused by benzodiazepines may also be associated with high-risk sexual behaviour including casual sexual contacts and unprotected sexual activity which appears to be a particular feature of temazepam abusers (Klee et al. 1990). Cognitive impairment, including deficits in learning and memory and in sustained attention, has been shown in many studies of long-term benzodiazepine users, even at therapeutic dose levels, and may persist after benzodiazepine withdrawal (Golombok et al. 1988; Tata et al. 1994; Gorenstein et al. 1995). Risks of maternal use during pregnancy include foetal developmental abnormalities (Laegreid et al. 1992), 'floppy infant syndrome' (Stirrat 1976) and a neonatal benzodiazepine withdrawal syndrome (Levy and Spino 1993).

Regular use of benzodiazepines, especially in high doses, readily leads to physical dependence, evidenced by withdrawal symptoms on sudden cessation (see below).

Many of the risks for injecting benzodiazepine users are common to self-injectors of all types of drugs (Table 15.2). It has been claimed that the use of temazepam is especially associated with the practice of sharing injecting equipment, thus increasing the risk of HIV infection and hepatitis (Klee et al. 1990). In addition, benzodiazepines, particularly temazepam (whether obtained from capsules, tablets or elixir), are extremely irritating and likely to cause tissue damage. When arm veins become occluded due to local irritation, users may proceed to injecting in the groin, where inadvertent intra-arterial injection has led to amputation. The severity of the addiction which can develop to temazepam is illustrated by the case of a temazepam injector who needed his leg amputated but was later admitted

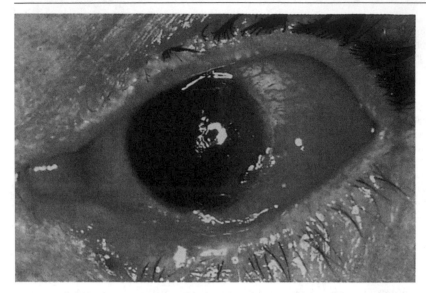

Figure 15.1 A complication of temazepam injection. A man aged 40 who misused drugs and had had a leg amputated after ischaemic damage from intra-arterial injections presented with blindness of recent onset. He was blind in both eyes. The left eye was ophthalmoplegic, with corneal clouding and no pupillary reflexes. This was the result of his injecting gel temazepam into the inner canthus. This substance is known to cause vascular occlusion (with permission from Thompson *et al*. 1993).

for a second amputation since he had continued injecting into his remaining leg (Parrott 1995). A second subject, following a leg amputation, injected temazepam gel into his eye, resulting in bilateral blindness (Figure 15.1).

Dependence and withdrawal symptoms

Dependence and a withdrawal syndrome has been well described in chronic benzodiazepine users over a range of therapeutic and supratherapeutic doses. There is less information on dependence and withdrawal in benzodiazepine abusers, who sometimes use the drugs intermittently or in binges and who often use other drugs. However, in a careful study, Seivewright and Dougal (1993) recorded symptoms in thirty-three polydrug users who for various reasons had abruptly stopped their benzodiazepines but not their other drugs. The results, shown in Table 15.3, clearly showed that the high-dose abusers experienced a profile of withdrawal symptoms similar to, though more severe than that described in low-dose users (Tyrer *et al*. 1990). The severity of symptoms was significantly related to the size of previous dosage in the abusers. It was notable in this study that the abusers had only stopped their benzodiazepines for 2–3 days, probably

Table 15.3 Symptoms reported during abrupt benzodiazepine withdrawal in thirty-three high-dose benzodiazepine and polydrug abusers

Depression
Shaking
Feeling unreal
Appetite loss
Muscle twitching
Memory loss
Motor impairment
Nausea
Muscle pains
Dizziness
Apparent movement of still objects
Feeling faint
Noise sensitivity
Light sensitivity
Peculiar taste
Pins and needles
Touch sensitivity
Sore eyes
Hallucinations
Smell sensitivity

before withdrawal symptoms had reached their peak. Even so, four of the subjects reported epileptic fits and there were sixteen reports of hallucinations and/or paranoid ideation. These severe symptoms may be especially common in high-dose abusers and have been reported in other studies (Stark *et al.* 1987; Busto and Sellers 1991; Scott 1990). Seivewright and Dougal (1993) noted that benzodiazepine abusers reported extreme anti-social behaviours in obtaining continued supplies, avoiding disruption of supplies and in drug-seeking behaviour when withdrawing. The severity and specificity of the benzodiazepine withdrawal syndrome was typified by one patient's remark: 'I'd rather withdraw off heroin any day. If I was withdrawing from benzos you could offer me a gram of heroin or just 20 mg of diazepam and I'd take the diazepam every time – I've never been so frightened in my life' (Seivewright and Dougal 1993: 20).

The incidence of withdrawal symptoms in benzodiazepine abusers is not known and deserves further study. Clinical observations suggest that some polydrug users, like some prescribed dose patients, can withdraw benzodiazepines without difficulty, especially polydrug users maintained on methadone. Regular daily benzodiazepine users are more likely than intermittent users to develop symptoms (Williams *et al.* 1996). The duration of withdrawal symptoms is also not clear; acute symptoms in the first few weeks may merge into prolonged anxiety and insomnia which may continue for weeks or months (Smith and Landry 1990).

Management of benzodiazepine withdrawal

Withdrawal methods for long-term prescribed therapeutic dose benzodiazepine users are well established and consist mainly of slow dosage tapering over weeks or months in an outpatient setting, combined with psychological support (Lader and Morton 1991; Ashton 1994). These methods are not entirely appropriate for high-dose benzodiazepine abusers. First, benzodiazepine abuse is often part of polydrug abuse and attention also has to be given to the primary drug. Second, a long period of outpatient dosage tapering is unlikely to be adhered to since additional benzodiazepines may be obtained illicitly. On the other hand, as described above, benzodiazepine abusers commonly use high doses and may be at particular risk of severe withdrawal symptoms including epileptic fits if the drugs are stopped abruptly. Therefore, a moderately rapid, controlled schedule of detoxification in an inpatient unit is preferable.

Several methods have been described. The most common technique is substitution of a slowly eliminated benzodiazepine (usually diazepam) for the abused drug (commonly, a shorter acting drug such as temazepam) followed by dosage tapering over 2 or more weeks (Harrison et al. 1984; Scott 1990; Williams et al. 1996). Some workers (Ries et al. 1989) have advocated the use of carbamazepine as an anticonvulsant in benzodiazepine withdrawal, though its ability to prevent other withdrawal symptoms is doubtful (Ashton 1994). A third method of detoxification recommended in some centres, particularly in the US, is phenobarbitone substitution (Smith and Landry 1990).

Longer term outpatient management of benzodiazepine abusers is problematic. Some centres have found benzodiazepines to be helpful in reducing overall illicit drug use and injecting (Greenwood 1996), and, occasionally, benzodiazepine use may be adaptive, allowing the opiate abusers to manage on a smaller dose of methadone (Strang et al. 1993). However, the overwhelming advice is that it is generally inadvisable to prescribe benzodiazepines as maintenance treatment for drug misusers (Smith and Landry 1990; Department of Health 1991, 1999; Seivewright and Dougal 1992; Seivewright et al. 1993) or alcoholics (see Case 15.2). A possible approach for opiate addicted patients who use benzodiazepines to increase the euphoriant effects of methadone is to alter the methadone treatment so that individuals feel less need for benzodiazepines (Seivewright et al. 1993). Alternatively, avoidance of benzodiazepines can form part of a contract for receiving a higher dose of methadone or other rewards (Stitzer et al. 1982). Nevertheless, it is clear that many benzodiazepine abusers, like prescribed users, suffer from anxiety, insomnia and depression. It may be that a flexible, individually tailored approach to benzodiazepine and other psychotropic drug prescribing, as well as psychological methods where practical, carried out in specialist centres, would bring the best results.

Unfortunately, in the present climate the rate of relapse after short-term benzodiazepine detoxification may be as high as it is with opiate detoxification (i.e. over 90 per cent after 1 year: Seivewright *et al.* 1993), and further experience is needed to establish the optimal long-term management. Meanwhile, efforts to reduce inappropriate prescribing of benzodiazepines both in general practice (Buetow *et al.* 1996) and in hospital (Zisselman *et al.* 1996) may help to decrease the quantity of benzodiazepines at present spilling into the illicit drug market.

References

Ashton, H. (1994) 'The treatment of benzodiazepine dependence', *Addiction*, **89**, 1535–41.
Bandolier (1998) 'Benzodiazepines and crashes', *Bandolier*, **5**(11), 6.
Borg, S., Carlsson, S. and Lafolie, P. (1993) 'Benzodiazepine/alcohol dependence and abuse', in C. Hällström (ed.) *Benzodiazepine Dependence*, pp. 119–27. Oxford: Oxford Medical Publications.
Buetow, S. A., Sibbald, B., Cantrill, J. A. and Halliwell, S. (1996) 'Prevalence of potentially inappropriate long term prescribing in general practice in the United Kingdom, 1980–95: Systematic literature review', *Brit. Med. J.*, **313**, 1371–4.
Busto, U. and Sellers, E. M. (1991) 'Anxiolytics and sedatives/hypnotics dependence', *Brit. J. Addict.*, **86**, 1647–52.
Carnwath, T. (1993) 'Temazepam tablets as drugs of misuse', *Brit. Med. J.*, **307**, 385–6.
Cole, J. O. and Chiarello, R. J. (1990) 'The benzodiazepines as drugs of abuse', *J. Psychiat. Res.*, **24**, 135–44.
Department of Health (1991) *Drug Misuse and Dependence: Guidelines on Clinical Management.* London: Her Majesty's Stationery Office.
Department of Health (1999) *Drug Misuse and Dependence: Guidelines on Clinical Management.* London: Her Majesty's Stationery Office.
Drug and Therapeutics Bulletin. (1997) 'Helping patients who misuse drugs', *DTB*, **35**, 18–22.
Evans, S. M., Funderberk, F. S. and Griffiths, R. R. (1990) 'Zolpidem and triazolam in humans: Behavioral and subjective effects and abuse liability', *J. Pharm. Exp. Ther.*, **255**, 1246–55.
Farrell, M. and Strang, J. (1988) 'Misuse of temazepam', *Brit. Med. J.*, **297**, 1402.
Golombok, S., Moodley, P. and Lader, M. (1988) 'Cognitive impairment in long-term benzodiazepine users', *Psychol. Med.*, **B18**, 365–74.
Gorenstein, C., Bernik, M. A., Pompéia, S. and Marcourakis, T. (1995) 'Impairment of performance associated with long-term use of benzodiazepines', *J. Psychopharmacol.*, **9**, 313–18.
Greenwood, J. (1996) 'Six years' experience of sharing the care of Edinburgh's drug users', *Psychiat. Bull.*, **20**, P8–11.
Greenwood, J. (2000) 'Stigma: Substance misuse in older people', *Geriatric Med.*, April, 43–5.

Griffiths, R. R. and Wolf, B. (1990) 'Relative abuse liability of different benzodiazepines in drug abusers', *J. Clin. Psychopharmacol.*, **10**, 237–43.

Griffiths, R. R., McLeod, D. R., Bigelow, G. E., Liebson, I. A. and Roache, J. D. (1984) 'Relative abuse liability of diazepam and oxazepam: Behavioral and subjective dose effects', *J. Psychopharmacol.*, **84**, 147–54.

Harrison, M., Busto, U., Naranjo, C. A, Kaplan, H. L. and Sellers, E. M. (1984) 'Diazepam tapering in detoxification for high-dose benzodiazepine abuse', *Clin. Pharmacol. Ther.*, **36**, 527–33.

Klee, H., Faugier, J., Hayes, C., Coulton, T. and Morris, J. (1990) 'AIDS-related behaviour, polydrug use and temazepam', *Brit. J. Addict.*, **85**, 1125–32.

Lader, N. and Morton, S. (1991) 'Benzodiazepine problems', *Brit. J. Addict.*, **B86**, 823–8.

Laegreid, L., Hagberg, G. and Lundberg, A. (1992) 'Neurodevelopment in late infancy after prenatal exposure to benzodiazepines – a prospective study', *Neuropediatrics*, **23**, 60–7.

Levy, M. and Spino, M. (1993) 'Neonatal withdrawal syndrome: Associated drugs and pharmacologic management', *Pharmacotherapy*, **13**, 202–11.

McNamee, D. (1994) 'Snorting flunitrazepam', *Lancet*, **344**, 187.

Mason, D., Birmingham, L. and Grubin, D. (1997) 'Substance use in remand prisoners: A consecutive case study', *Brit. Med. J.*, **315**, 18–31.

Moos, R. H., Nichol, A. C. and Moos, B. S. (2002) 'Risk factors for symptom exacerbation among treated patients with substance use disorders', *Addiction*, **97**, 75–85.

Parrott, A. (1995) 'State of the art. Psychoactive drugs of use and abuse: Wobble, rave, inhale or crave?', *J. Psychopharmacol.*, **9**, 390–1.

Pedersen, W. and Lavik, N. J. (1991) 'Adolescents and benzodiazepines: Prescribed use, self-medication and intoxication', *Acta Psychiatrica Scand.*, **84**, 94–8.

Ries, R. K., Roy-Byrne, P. P. and Ward, N. G. (1989) 'Carbamazepine treatment for benzodiazepine withdrawal', *Amer. J. Psychiat.*, **146**, 536–7.

Robertson, J. R. and Ronald, P. J. M. (1992) 'Prescribing benzodiazepines to drug misusers', *Lancet*, **339**, 1169–70.

Ruben, S. M. and Morrison, C. L. (1992) 'Temazepam misuse in a group of injecting drug users', *Brit. J. Addict.*, **87**, 1387–92.

Scott, R. T. A. (1990) 'The prevention of convulsions during benzodiazepine withdrawals', *Brit. J. Gen. Prac.*, **40**, 261.

Seivewright, N. and Dougal, W. (1992) 'Benzodiazepine misuse', *Curr. Opin. Psychiat.*, **5**, 408–11.

Seivewright, N. and Dougal, W. (1993) 'Withdrawal symptoms from high dose benzodiazepines in polydrug users', *Drug Alcohol Depend.*, **32**, 15–23.

Seivewright, N., Donmall, M. and Daly, C. (1993) 'Benzodiazepines in the illicit drugs scene: The picture and some treatment dilemmas', *Int. J. Drug Policy*, **4**, 42–8.

Serfaty, M. and Masterton, G. (1993) 'Fatal poisonings attributed to benzodiazepines in Britain during the 1980s', *Brit. J. Psychiat.*, **163**, 386–93.

Shaw, M., Brabbins, C. and Ruben, S. (1994) 'Misuse of benzodiazepines', *Brit. Med. J.*, **308**, 1709.

Smith, D. E. and Landry, M. J. (1990) 'Benzodiazepine dependency discontinuation: Focus on the chemical dependency detoxification setting and benzodiazepine-polydrug abuse', *J. Psychiat. Res.*, **24**, P145–56.

Stark, C., Sykes, R. and Mullen, P. (1987) 'Temazepam abuse', *Lancet*, **October**, 802–3.

Stirrat, G. M. (1976) 'Prescribing problems in the second half of pregnancy and during lactation', *Obstet. Gynecol. Surv.*, **31**, 1–7.

Stitzer, N. L., Bigelow, G. E., Liebson, I. A. and Hawthorne, J. W. (1982) 'Contingent reinforcement for benzodiazepine-free urines: Evaluation of a drug abuse treatment intervention', *J. Appl. Behav. Anal.*, **15**, 493 503.

Strang, J. (1984) 'Intravenous benzodiazepine abuse', *Brit. Med. J.*, **289**, 964.

Strang, J. (1991) 'Abuse of buprenorphine (Temgesic) by snorting', *Brit. Med. J.*, **302**, 969.

Strang, J., Seivewright, N. and Farrell, M. (1993) 'Oral and intravenous abuse of benzodiazepines', in C. Hällström (ed.) *Benzodiazepine Dependence*, pp. 128–42. Oxford: Oxford Medical Publications.

Strang, J., Griffriths, P., Abbey, J. and Gossop, M. (1994) 'Survey of use of injected benzodiazepines among drug users in Britain', *Brit. Med. J.*, **308**, 1082.

Tata, P. R., Rollings, J., Collins, M., Pickering, A. and Jacobson, R. R. (1994) 'Lack of cognitive recovery following withdrawal from long-term benzodiazepine use', *Psychol. Med.*, **24**, 203–13.

Thompson, J. L., Honeybourne, D. and Ferner, R. E. (1993) 'Minerva', *Brit. Med. J.*, **307**, 1434.

Tyrer, P., Murphy, S. and Riley, P. (1990) 'The benzodiazepine withdrawal symptom questionnaire', *J. Affect. Disord.*, **19**, 53–61.

Warburton, D. M. (1988) 'The *in vitro* pharmacology of selective opioid ligands', in M. Lader (ed.) *The Psychopharmacology of Addiction*, pp. 27–49. Oxford: Oxford University Press.

Wilkinson, P. (1997) '"Rat boy" given four years for toby jug theft', *The Times*, 4.

Williams, H., Oyeffeso, A. and Ghodse, A. H. (1996) 'Benzodiazepine misuse and dependence among opiate addicts in treatment', *Irish J. Psychol. Med.*, **13**, 62–4.

Woods, J. H., Katz, J. L. and Winger, G. (1988) 'Use and abuse of benzodiazepines. Issues relevant to prescribing', *J. Amer. Med. Assoc.*, **260**, 3476–80.

Zawertailo, L. A., Busto, U., Kaplan, H. L. and Sellers, E. M. (1995) 'Comparative abuse liability of sertraline, alprazolam, and dextroamphetamine in humans', *J. Clin. Psychopharmacol.*, **15**, 117–24.

Zisselman, N. H., Rovner, B. W. and Shmuely, Y. (1996) 'Benzodiazepine use in the elderly prior to psychiatric hospitalization', *Psychosomatics*, **37**, 38–42.

Chapter 16

Treatment options across the addictions

John McCartney

Key points

The focus of this chapter is to understand the reasoning behind treatment choices. It does not offer a 'recipe' for treatment, which would only fit a particular time and place. It does aim to promote decision-making skills which can be applied wherever the problems of dependence are encountered:

- *Treatment methods are influenced by causal beliefs. The biopsychosocial view fits the evidence most adequately at the current time.*
- *Although assessment and treatment overlap, a comprehensive, multidisciplinary evaluation or assessment is generally required before medical or psychosocial options can safely be selected and integrated into a coherent treatment plan.*
- *Possible medical interventions and pharmacotherapy are described. These methods tend to parallel psychosocial approaches, at least in more intensive treatment settings.*
- *Selected psychological and psychosocial options are described and illustrated, with an emphasis on systems thinking.*
- *Treatment evaluation supports the use of different methods, with an emphasis on matching according to client characteristics and needs. The optimal, coherent integration and differentiation of the various options is a crucial issue for clinicians and researchers to investigate in more depth.*

Introduction to treatment approaches

A close bond exists between beliefs concerning the causes of addiction and actions towards those who experience such difficulties. Moral condemnation, attempts to impose legal sanctions, religious persuasion and medical (biological) treatments have all been used in the past because their use was implied by the dominant belief system or *zeitgeist*. In more recent times, although the medical–psychiatric view has tended to predominate, social and psychological causes have been given more attention, altering the way addicts are 'managed' or 'treated' and resulting in a *biopsychosocial* (i.e. biological–psychological–social) perspective. Whilst being integrative, this view retains respect for variability amongst drug users, especially regarding their own beliefs about causality, therefore also serving to *differentiate* between individuals in a helpful manner (Lindstrom 1992).

Prochaska and DiClemente (1982) discuss a trans-theoretical framework useful in differentiating broad groups of drug and alcohol users in terms of their place in what they label the 'cycle' or 'stages' of change (see chap. 10, fig. 2). The focus in their model moves from causality to the more practical, and relevant, differences between addicts in their attitude to change and their varying hopes, expectations, needs and fears. If patients or clients[1] have not seriously considered change, they are thought to be in the '*pre-contemplation*' stage. Individuals sometimes attend treatment services because they feel forced to seek help and not necessarily because they want to do so. Where they have not yet made a *decision* to change, it would be foolhardy to attempt treatment in advance of testing and exploring their motivation in more depth. Those who have decided to change, either towards abstinence or control, are considered to be in the '*initiation stage*' and usually need help in establishing a period of abstinence.[2] If clients succeed in drug use reduction or cessation, they enter the '*maintenance stage*', where they concentrate on learning techniques to sustain change and avoid relapse. When clients maintain change for an appreciable time, they begin to break out of the tendency to repeat the maladaptive addictive 'solution'. If unsuccessful, they may relapse and the change cycle begins again, although perhaps facilitated by prior learning and experience. Whilst the cycle emphasizes individual factors, context is vital to bear in mind, because personal success or failure tends to be strongly influenced by pretreatment characteristics including the strengths (e.g. good relationships, work, social support) and weaknesses (e.g. stressors, unemployment) in the social system (Moos *et al.* 1990).

Venues, context and setting

The main variables differentiating treatment agencies, or services, include the *models* or *theories* workers hold to explain dependency behaviour, the *resources* available to the service and the relative *intensity* or *depth* of approach. Staff are the main resource and they can be medically oriented

---

1 These words are used interchangeably in this chapter.
2 In the rare cases where controlled drinking is the goal, a substantial period of abstinence is essential.

(e.g., psychiatrists, psychiatric or community psychiatric nurses), non-medical (e.g. social workers, clinical psychologists, counsellors) or, perhaps more commonly, a balance between the two.

Table 16.1 summarizes the main services in relation to treatment intensity and possible venues. Liaison services refer to links between addiction treatment agencies and other services where individuals have medical or social problems resulting, at least in part, from alcohol and drug misuse. Minimal intervention services, including general practitioner treatment, orientate themselves to successful detoxification, usually by providing a temporary substitute like oral methadone instead of injected heroin. These, less intensive, approaches help patients cope with the effects of withdrawal and can encourage addicts to engage in more intensive treatment, where necessary (Kleber 1989). More vulnerable patients may require drug dependency unit or specialist outpatient services. The most medically vulnerable, and the socially unstable, generally need residential and inpatient settings. These venues provide greater protection from environmental cues and pressures to continue using or relapse and can help patients develop more effective coping styles and interpersonal behaviours. Other possible treatment venues include private and voluntary services.

Although there are these differences between venues and professions, a common factor in most interventions is the need for a good therapeutic relationship. A beneficial alliance is manifest in a containing structure: a relationship and setting having the potential to share and hold the patient's positive and negative expectations. Drug clients tend to lack an internal capacity to bear, and appropriately express, emotion and conflict, evident in mood swings and impulsivity, resulting in low self-esteem. These problems are partly a cause and partly a consequence of substance misuse. The development of communication skills and a self-containing structure in the patient is only gradual and requires workers to manifest a forbearing attitude, genuine warmth and creative constructiveness.

It is generally impractical to provide a long-term continuous relationship in most clinical settings, therefore the *focus* in treatment is often on helping patients develop more constructive attitudes and to utilize *facilitating* environmental structures in order to sustain effort, achieve change and re-inforce abstinence. These environmental structures can include improvements in employment or relationships, new relationships, support groups, probation officers, religious or spiritual support, AA and NA (Vaillant 1988). A modified *lifestyle* is necessary to develop alternative reinforcements and improved social functioning.

Evaluating drug and alcohol problems

The aim of evaluation is to appraise carefully the patient's needs in order to negotiate an optimal treatment response. This process requires a holistic

Table 16.1 Addiction treatment services described in relation to intensity and possible venues or locations

Intensive interventions	Varied intensity interventions	Minimal interventions	Liaison services	Non-statutory services
Residential therapeutic communities	Drug dependency units and outpatient clinics	GP detoxification	Medical wards	Work settings (employee assistance programmes)
Inpatient services	Day care services	Needle and syringe exchanges	Accident and Emergency departments	Alcoholics Anonymous
Rehabilitation units	Community teams	Outreach services	Prison/Probation services	Narcotics Anonymous
Hostels and halfway houses	Street agencies	Telephone advice helplines	Social services departments	Volunteer counselling services
	HIV/AIDS and health counselling services			

perspective where physical, social and psychological health are given equal attention. A *multidisciplinary* approach is, therefore, especially relevant to the addictions.

The first stage in evaluation involves engagement and a systematic review of the individual's:

- drug use;
- physical health;
- life history; and
- treatment expectations.

In medical settings, the *Diagnostic and Statistical Manual* (DSM-IV) can be used in order to identify substance abuse and dependence. The main diagnostic signs include amount used, relative frequency, life interference and, where relevant, tolerance and withdrawal symptoms. Initial interviews are sometimes difficult to conduct if clients are primarily concerned with the physical effects of stopping substance misuse. They can also have forensic problems. Careful attention is needed to possible medical complications, in particular. Severe withdrawal symptoms (e.g., after alcohol or barbiturates) can be life-threatening and, therefore, may require specialist medical attention, discussed later in this chapter.

An accurate assessment of the level or *degree of dependency* is always essential (even in non-medical settings), perhaps best quantified by a combination of verbal accounts, collateral report, questionnaires and objective indexes (like urine screening). When drug use has been satisfactorily quantified, the assessment can focus on the more qualitative understanding of alcohol and drug intake, utilizing a functional view. This involves identifying the most personally relevant *antecedents* and *consequences* of use and abuse (Davies 1992). Some clients find it difficult to clarify the thoughts, emotions and events surrounding use and may need to keep a *diary* to succeed at this task. The idea of the substance as a panacea or source of intense enjoyment is often more illusory than real when the actual (in contrast to the imagined) consequences and effects are more completely recognized.

If clients are *intravenous* drug users, HIV/AIDS screening and counselling is normally necessary. Legal matters in relation to illicit drugs should be clarified with the patient (London and Ghodse 1989) and, in the UK, patients should be informed of the treatment agency's contribution to a regional drug database. Urine screening is essential before any prescribing and tends to facilitate rather than jeopardize a trusting relationship.

Amidst all the practical concerns relating to patient screening and assessment, drug workers can sometimes neglect the *emotional task* involved in engaging the patient's active and behavioural co-operation in change. Clients tend to feel a lack of control and low self-esteem and need help in expressing their hopes and fears. Oppenheimer *et al.* (1988) found that the main anxieties in drug misusers relate to the prospect of failure, rather than success, and the fear of having controls, rules or punishments unilaterally imposed on them. Although addicts often fear rejection, they may appear angry instead of anxious, perhaps feeling more ashamed or guilty than they consciously realize. Emotional contact can be strengthened if workers, judiciously and in good time, label these, and other, affects.

A family assessment is often useful and is mandatory where substance misuse may have interfered with childcare or where violence and anger may be uncontained. Tact is required because this may appear threatening to already ambivalent patients. Clear institutional policies with respect to childcare are essential in order to help workers take effective action to protect children and to link well with social services departments and health visitors in primary care settings. Family or work stress can be a cause or consequence of drug use. Perhaps, more commonly, a vicious cycle is evident whereby initial tensions or difficulties between individuals have precipitated or reinforced drug use and the 'solution' (drug use) has itself become the more apparent problem (rarely the only problem!). In negating or suppressing their self-consciousness, substance dependent clients tend to neglect the impact of their behaviour on the family or work *system.*

A test for your understanding thus far

Based on a study of alcohol use disorders in Birmingham (Commander *et al.* 1999), 560 people with drink problems were detected in one district. Of these people living in the community, 287 took some problem to primary health care teams, of whom 122 were then recognized as having an alcohol-related disorder, of whom 13 received a referral to any specialist service, of whom only 1 was ever admitted for specialist inpatient care.

- How typical of cases living in the community do you think the patients seen in an inpatient addiction unit are?
- What do you think are the most crucial differences between those thought to require specialist care and those thought not to require this kind of help?

Structuring treatment options

The evaluation or assessment should yield information relating to the patient's needs, personality and social circumstances, in addition to their psychological and medical health. The stage reached in the change cycle and the strengths and weaknesses in the client's social system should be evident. The assessment should, in essence, identify the causal factors and help specify *what* needs to change in order for the person to successfully alter their substance dependency (e.g. coping strategies, improved relationships). The most important blocks, or perceived blocks, to change should also be clear. This results in a treatment contract, including time limits, goals, the planned options themselves and identification of possible risks. Contracts help avoid inertia or drift, increase uptake of personal responsibility and encourage the respect for rules and boundaries needed to attain change.

The biopsychosocial model implies drug problems have a variety of causes and consequences. Therefore, in response to this realization, various psychological and psychosocial interventions tend to be used *in parallel* with medical treatment, although, sometimes, patients need first to stabilize on substitutes before they can benefit from such help. Table 16.2 summarizes the main treatment options available to treatment services. The specific options selected depend on client needs and venue. Linking to Table 16.1, 'minimal intervention services' restrict themselves to advice and medical treatment. More intensive interventions (including most non-statutory services) utilize a wider range of psychosocial approaches. Commitment and personal responsibility is strengthened if clients are actively involved, in an informed way, in selecting the options best suited to helping them achieve secure change. Where individuals have characteristics indicative of good outcome – including lower levels of dependence on only one substance, stable employment, protective environments and adaptive coping strategies – simple, accurate and appropriate *advice* may well be sufficient. At the least, treatment can alleviate the vicious cycle of misuse (Gawin and Ellinwood 1988). Where drug problems are more severe, or the social system less intact, more intensive work, including inpatient treatment, may be required (discussed also in Chapter 17).

After *what* needs to change has been adequately clarified, treatment can be conceptualized in terms of two main dimensions, one relating to motivation for change and the other to strategies for achieving and maintaining change.

Why – Motivation

Patients often see this as having to do with the presence or absence of reasons for change. Motivation is crucial because it relates to the energy available for effort directed at change and can gradually be strengthened if the patient is adequately *engaged* in the treatment work. Almost 5,000 people entering a wide variety of care settings were followed up by the US Drug Abuse Treatment Outcome Study (Joe *et al.* 1998). Simple

Table 16.2 Possible treatment choices or options in addiction treatment

Medical interventions	Psychological treatment	Systems interventions	Group treatment	Alternative therapies
Diagnosis	Advice	Marital therapy	Relapse prevention	Religious–spiritual support and guidance
Liver function and other biochemical testing	Counselling	Family therapy	Interpersonal group therapy	Creative therapies (e.g. art, music, drama)
Harm minimization	Motivational interviewing	Workplace liaison	Awareness groups	Occupational therapy
Detoxification	Behavioural techniques	Network linking	Self-help and support groups	Physical therapies
Maintenance contracts	Cognitive therapy	Consultancy work (e.g. advice on company alcohol policies)	Family support groups	Complementary medicine
Antagonists (blockades)	Psychotherapy		NA, AA	Training in leisure activities
Psychotropic medication				

measures of their engagement (like the degree to which a client 'was working on his/her problems') predicted very well whether a client would stick with any treatment programme. Motivational interviewing can help identify and reinforce desire for change. This technique is discussed on p. 225 and also in Chapter 10.

How – method or strategy

Managing, or facilitating, change involves helping clients identify a strategy to achieve and sustain abstinence. This sometimes requires medical support, often benefits from improved social systems and usually requires behavioural change. Medical treatment, advice, behavioural–cognitive therapy, psychotherapy and systems work are, therefore, all possible elements in *How* to achieve change. They are means to an end and not ends in themselves. This is important to clarify in pharmacotherapy, because alternative drugs are sometimes construed as being a reason to change instead of a means to attain change. Such a misapprehension is perhaps understandable in that, whilst distinct, the two main change dimensions overlap in the sense that when the person feels more able to adapt to life without drugs, through the various medical and non-medical approaches, motivation itself may increase. Confusion is less likely if their conceptual separateness is borne clearly in mind.

Selected medical interventions and treatment

The main medical interventions are summarized in Table 16.2.

Medical diagnosis

Medical or nursing evaluation and intervention is often essential in order to diagnose possible physical sequelae of substance abuse (e.g. liver damage, HIV infection, hepatitis), in addition to predicting and alleviating withdrawal symptoms. Recognition of sustained or imminent physical harm can be a strong motivator to engage in behaviour change. Psychiatrists use structured aproaches to diagnose substance abuse and dependency (e.g., DSM-IV), and to help identify psychiatric symptoms and sequelae, including depression, suicidal tendencies and drug-induced psychotic symptoms. These problems may be causes or consequences of drug use and neuropsychiatric problems, in particular (e.g. specific memory and cognitive problems), may be due to HIV or AIDS (Department of Health 1991).

Pregnant addicts, and especially pregnant alcohol misusers, require very careful evaluation and liaison with obstetrics and gynaecology. Although all drugs may have harmful effects on the developing embryo, excessive alcohol consumption has been shown to cause foetal abnormalities (foetal alcohol syndrome). Pregnant intravenous drug users and those who are already HIV-positive need specialist medical attention.

Harm minimization

A valuable health education opportunity is available to patients when they are reviewed during the medical and nursing assessments. Although patients may be reluctant to change their drug or alcohol dependency, at least they may benefit from informed advice concerning drug-related behaviours. The main physical health risk in illicit drug use is linked to injecting behaviour and not drugs themselves. Advice on safer practices, hepatitis B immunization, attendance at needle and syringe exchanges and similar clinical care can prevent harm to the patients and can help stop the spread of serious infections, including HIV and hepatitis. Health education advice on lifestyle, diet and other medical matters can increase motivation to change addiction behaviour by clarifying the serious risks and disadvantages in continued dependency.

Detoxification and maintenance

Table 16.3 is a brief summary of the main types of pharmacotherapy, providing examples of possible drug treatments and indicating their main use, as well as probable benefits and possible hazards (see Kleber 1989; Ghodse and Maxwell 1990; Department of Health 1999 for additional examples, specific detoxification regimes and more in-depth discussion). Drugs can be essential in preventing or alleviating unpleasant and, at times, life-endangering withdrawal reactions, or they may be optional when the problem is more to do with motivation, volition and mood. Patients with a prior history of withdrawal complications, and those who are highly dependent, need particularly careful medical evaluation. Urine screening normally precedes prescribing and regular monitoring is needed in order to identify relapse.

Pharmacotherapy is not always necessary or desirable. The popular stereotype (in patients and workers alike) implying medical doctors are employed mainly to prescribe drugs is a misperception that needs to be challenged. Drugs having a consistent withdrawal syndrome include opioids, alcohol, tobacco and benzodiazepines. However, other substances (e.g. cannabis) may not produce such physical effects. Even where withdrawal symptoms are evident, they may not be sufficiently serious to warrant pharmacotherapy. Where administered, detoxification or stabilization is generally seen 'as a preliminary step' (Gossop et al. 1986) in treatment and in developing a substance-free adaptation to life. Substitute drugs can help to contain patients' distress in order to facilitate their engagement in

Table 16.3 Selected examples to illustrate possible drug treatments in managing substance abuse and dependency

Dependency problem	Drug treatment	Main use(s)	Potential benefits	Potential hazards
Opioid/Opiate dependence	**Methadone** (usually oral in contrast to injections)	Relieves opioid withdrawal symptoms. Longer term maintenance substitute	Decreases illicit drug use. Harm minimization. Adjunct to psychological therapy	Mainly transitory, including sweating, constipation and amenorrhoea. Possible serious complications if combined with other drugs. Possible iatrogenic dependence
	Naltrexone Naltrexene	Opioid antagonists or 'blockades' Naloxone is used to reverse narcotic overdose	Non-addictive. Relapse deterrent Promotes sustained abstinence. Improves treatment retention rate.	Variable compliance and effectiveness Can precipitate severe withdrawal symptoms if used with patients who are not opioid-free. Attempts to overcome the blockade with high doses of opioids can be life-threatening
Alcohol dependence	**Benzodiazepines**	Control or relief of alcohol withdrawal symptoms. Short-term adjunct for severe anxiety symptoms. Rapid opioid withdrawal regimes	Minimizes the main physical alcohol withdrawal symptoms (e.g. convulsions, DTs). Diminishes subjective and physical anxiety/apprehension	Substitute dependency if not used within treatment guidelines. Potentially lethal if combined with alcohol
	Chlormethiazole ('heminevrin') **Disulfiram** ('antabuse') **Acamprosate**	Treatment of alcohol withdrawal symptoms. Anticonvulsant Results in unpleasant symptoms if alcohol is ingested	Minimizes alcohol withdrawal symptoms Relapse deterrent	Danger of respiratory depression. More addictive than benzodiazepines If alcohol is ingested, purgative effects can be dangerous, occasionally causing cardiovascular collapse
Tobacco dependence	**Nicotine** (patches, chewing gum inhalers, nasal spray)	Short-term substitute when withdrawing	Relieves withdrawal symptoms, if used with psychological help	Dependency on the substitute

Drugs used to treat more specific withdrawal symptoms:

Dependency problem	Drug treatment	Main use(s)	Potential benefits	Potential hazards
Opiate/Opioid dependence	**Lofexidine Clonidine**	Relief of the physiological symptoms of withdrawal	Non-addictive	Can lower blood pressure and heart rate
Stimulant dependence (e.g. crack, amphetamines)	**Antidepressants** (tricyclics, SSRIs)	Relief of mood states linked to severe psychological dependence and consequent adverse withdrawal response	May prevent suicide and severe depressive illness. Can improve adherence to psychological treatment	If taken to excess, tricyclics can be cardiotoxic

Thanks go to Alex Baldacchino and Woody Caan for their constructive advice and recommended changes to this table.

psychosocial therapy. Medical and psychosocial interventions tend to be complementary rather than 'alternatives'.

The short-term use of drugs (e.g. methadone) to quell withdrawal symptoms should be differentiated from *maintenance* medication, where a contract is negotiated between the service and patient to gradually reduce at an agreed pace and/or stabilize using behaviour for a longer time period. A recent review by Bandolier (1999) found that the most common benefit from methadone maintenance was a reduction in drug-related criminal behaviour: 85 per cent of patients reduced their involvement in crime over 6–12 months. However, agencies should recognize that long-term prescribing may not always encourage patients to develop the needed motivation to learn to cope without drugs in healthier and more mature ways, through personal development and effort. Drug dependency may, inadvertently, be perpetuated beyond its natural course.

Pharmacotherapy interventions following, or adjunctive to, detoxification

Other drug treatments include use of 'antagonists'. These substances, including naltrexone, are a different group of drugs, administered when patients are drug-free. They block the euphoric effects of the problematic drug in the event of use, hence discouraging a relapse to heroin. Antagonists to opiates are also used after the controversial 'ultra-rapid detox' carried out under 6–8 hours of general anaesthesia (Álvarez and Del Río 1999). For patients who are newly drug-free, 'sensitizing agents' can cause very unpleasant or even dangerous side effects if they return to their former drug use. The sensitizing agent most widely used in alcohol dependence has been disulfiram, either as Antabuse tablets or as implants. However, a systematic review of disulfiram research by West (2000) found that it is not necessarily sufficient to sustain sobriety. More recently, drugs to reduce craving when abstinent have been investigated, although none are yet in widespread use. The best supported anti-craving agent for alcoholism seems to be acamprosate (Cornish and O'Brien 1999); for example, in a study of German alcoholics treated with acamprosate, 43 per cent were able to sustain abstinence compared with 12 per cent given a placebo. Psychotropic drugs can also be used to treat specific psychiatric symptoms, including mood stabilizers (e.g. carbamazepine), antipsychotics (e.g. thioridazine) and antidepressants. In amphetamine abuse, for example, whilst no physical withdrawal pattern may be evident, serious depression can lead to suicide risk. Some of these patients may therefore benefit from appropriate medication (e.g. those with a history of depression in self or family).

It should be borne in mind that not all medical treatments can be dispensed safely in outpatient settings. Morgan (1990) strongly recommends, in relation to alcohol dependency, that chlormethiazole (heminevrin) only be administered on an inpatient basis. Such settings may also be essential to

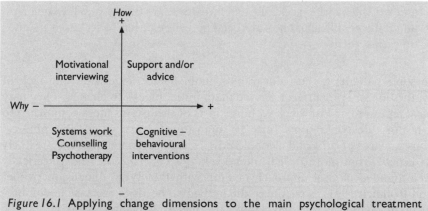

Figure 16.1 Applying change dimensions to the main psychological treatment options (McCartney 1996, reprinted with the publisher's permission). The horizontal axis refers to the patient's degree or level of motivation (+ corresponds to more, – corresponds to less). The vertical axis refers to beliefs about their capacity to change (e.g., knowledge of strategies, access to social support).

assess properly the patient's true level of dependency. Special care is needed in polydrug misuse because of the lethal, potentiating effects of certain combinations (e.g. alcohol and sedatives).

Selected psychological and psychosocial treatment options

The psychological therapy options, or components, summarized in Table 16.2, can be conceptualized in relation to the main change dimensions already discussed (see Figure 16.1). The horizontal axis relates to motivation for change and the vertical axis to the individual strategies and psychosocial or systems methods designed to facilitate behavioural change. In addition to developing motivation and coping skills, psychological methods help to instil self-control by improving the client's cognitive and behavioural rationality. Clear aims and specific goals are a prerequisite to successful treatment because these add structure, pattern and direction to clients' lifestyles, counteracting inertia detrimental to change. The options discussed here are only some of the possibilities (Table 16.2, columns 2–5). Often, more than one option is selected, either in sequence or parallel.

Motivational interviewing

A 44-year-old patient comes along to a drug and alcohol service visibly anxious and depressed. He admits to drinking up to one bottle of spirits a day and says he has experienced difficulties in his marriage

> and work. His wife, and colleagues at work, are (in his view) 'exaggerating' the problem and he's annoyed at feeling he has to seek treatment 'to please them'.

In some patients (as just illustrated), ambivalence, indecision and lack of motivation or commitment are the main initial blocks to change. They may have repressed or lost touch with the parts of themselves more committed to a healthy lifestyle and may also be unaware of actual or possible physical consequences of drug and alcohol abuse. These 'precontemplators' are likely to benefit from motivational interviewing, a conceptually and empirically well-founded clinical approach dealing with the 'Why' dimension (Miller and Rollnick 1991). Sometimes, the individual needs to increase in commitment to change before the 'How', the means to achieve change, can be adequately tackled because capacity for change tends to increase with conscious recognition of ambivalence, positive reasons for change and informed decision making. The case vignette illustrates a common conflict, wherein the client seems to experience desire for change coming from outside rather than inside.

Motivational interviewing avoids passing judgement, instead emphasizing the need to understand the *function and the effects* of substance use and misuse; what it does for and to the individual; how it facilitates or blocks adaptation and interaction. Such exploration is likely to increase thoughtfulness, personal control and responsibility, whilst minimizing resistance stimulated by an adversarial or punitive approach. Clients can begin to consolidate their commitment to attempt change when they begin to weigh rationally the pros and cons (for self and others) of continued use, in contrast to more healthy behaviours and lifestyles. A balance sheet approach can help the patient reach and implement firm and reasoned choice behaviour (McCartney 1997).

Cognitive–behavioural therapy

Behavioural and cognitive methods may be used (separately or in combination) if clinicians have training in such techniques (e.g. clinical psychologists, nurse behaviour therapists). In general, the strategies aim to increase self-control and problem-solving skills during the initiation and maintenance of change stages (Carroll 1999). Their use depends on *individualized* understanding of the *function* of addictive behaviour, conceptualized in terms of antecedents and consequences. The antecedents, or precursors, may be internal (feelings, thoughts, conflicts) or external (situational cues, interpersonal conflict); consequences may involve pleasure, on the one hand, or/and the avoidance of physical or psychological pain and discomfort, on the other. Identifying punishing effects can help encourage change, especially when damaging consequences for their system(s) are addressed.

'Relapse prevention' therapy (Scott 1989; Wanigaratne *et al.* 1990) combines behavioural and cognitive methods and involves a systematic attempt to identify dangerous situations or disturbing emotions, the objective being to help clients learn, implement and practise strategies to cope more effectively with conditioned cues, craving and urges to use. Cognitive methods include self-monitoring, self-control and decision-making. Behavioural techniques include social skills training, cue avoidance, cue exposure, relaxation and other coping methods. The goal is to help the client learn and practise a constructive, rewarding behaviour in direct opposition to the addictive tendency.

Cue avoidance involves the patient implementing their knowledge of dangerous situations by evading the stimuli linked to drug use and by learning to tolerate craving. Exposure, on the other hand, consists of introducing the patient, in a guided and gradual way, to difficult and unavoidable settings or experiences (e.g. drug-using friends), after helping them learn self-control strategies. In common with other behavioural methods, relapse prevention can be group or individual-based.

Systems approaches

A young 25-year-old woman seeks help from a drug dependency unit. She appears worried and upset, frightened at her increasing use of heroin and the fact she and her partner are finding it hard to abstain, despite lack of financial resources and their recently born first child. She stopped using illicit drugs during her pregnancy and this gives her confidence in her ability to regain abstinence. She experiences her husband's use of drugs as a source of constant temptation and anger.

There is a tendency to think in simplistic terms concerning the link between substance misuse and family or marital relationships, seeing the addict as either victim *or* persecutor. The systems perspective emphasizes the role of *interpersonal* and psychosocial causes and effects, in contrast to an artificial focus on individuals in isolation. Identifying *reciprocal* dysfunctional communication can help avoid possible counter-therapeutic reactions in the system, including overambitious attempts to 'rescue' and hostile rejecting responses. The systems view also encourages workers to consider the needs of those more vulnerable than the addict, especially children.

In this example, it is possible to hypothesize an interaction, or vicious cycle, between the patient's substance misuse and her partner's behaviour. Family, or marital, therapy (Galanter and Kleber 1994; Barth *et al.* 1993) can increase interpersonal awareness and help modify dysfunctional interactions, thus helping the system to develop new styles or patterns of interaction, reinforcing rather than blocking change. Various family approaches to managing dependence have been explored (O'Farrell and Fals-Stewart 1999), sometimes successfully. Addicts don't always register the harmful impact of their behaviour on the system and their motivation for change may increase if the hurt or resentment in others is recognized. In this vignette, the mother may feel alone with the child, unsupported and angry that her willingness to change is not matched by an equal effort on the part of her spouse. Family therapy practice requires training and supervision or specialist referral in its own right. Alternative help for spouses can be provided in family support groups. The Mental Health Foundation have developed a training programme for health visitors and practice nurses to improve social support and coping strategies for spouses and parents (Copello 2000). Therapy for the children may be provided individually, in family settings or in groups.

Systems thinking emphasizing functional relationships can also be applied more generally (Ghodse and McCartney 1992). Drug and alcohol agencies can try to link with the client's treatment network (e.g., GP, social worker) and their work setting, where relevant. Integrative linking leads to a coherent treatment response, which is necessary to help avoid mistaken or contradictory advice. Aftercare integrating community social workers and support groups with specialist alcohol and drug treatment programmes has been found to be generally helpful (White 1998).

Counselling

Addictions workers differentiate between 'keyworking' clients and counselling clients. Professional counselling is a strategic and structured approach to helping clients change and should not be considered the same as informal meetings or case co-ordination ('keyworking'). Counselling is especially relevant to the addictions because it can help clients contact and articulate emotional barriers to achieving and maintaining change. Drug and alcohol patients often have difficulty identifying meaning, emotions and conflicts accurately because the substances have reinforced affect control, denial and suppression. Workers can sometimes inadvertently discourage emotional contact, mirroring, instead of responding to, a common problem in clients themselves. Empathy, genuineness and warmth in the counsellor help strengthen the therapeutic relationship and increase the client's self-awareness and self-esteem, vital catalysts in successful change. Emotional understanding and containment can be balanced with supportive confrontation, where necessary.

Counselling work can be more or less specialized and more or less reflective, depending on the counsellor's training and the client's needs. Integrated counselling blends emotion-centred and behavioural methods, including coping advice, diary keeping, goal setting and similar client self-structuring strategies (Egan 1998). Specialized counselling and advisory services address more specific needs (e.g. health education, vocational guidance) or subgroups, including intravenous drug users. Intensive counselling methods overlap with psychotherapy in attempting to understand less conscious or unconscious meanings and conflicts. Psychoanalytic approaches are discussed elsewhere (e.g. Schneider and Khantzian 1992).

Treatment effectiveness

In common with other *biopsychosocial* disorders, the literature on the outcome of treatment is easier to oversimplify than summarize. Comparisons between therapies and the debate over intensive versus minimal interventions have been especially contentious issues (Miller 1998). Similarities in outcome across the various psychosocial and medical options are generally more striking than differences, although this is, in part, a consequence of formal and informal 'therapist–patient-treatment' matching (Allen and Kadden 1995) and a tendency to use options in combination. When selected options or components match the client's needs and are well integrated, their combination tends to have a synergistic, not a cancelling-out or destructive, effect. Neither medical nor psychological treatment benefits all patients, although effects should be perceived dimensionally, not dichotomously, with some clients partly, not completely, recovering. For example, change may be manifest in safer drug practices, or in controlled use (especially in alcohol dependent patients), not necessarily in total abstinence.

Whilst well-designed minimal interventions have found empirical support (Bien *et al.* 1993), heavily addicted patients tend to benefit more from intensive help, perhaps because these methods also address social and psychological needs. Hence, less intensive treatments are more safely employed when referral for specialist counselling or intensive treatment is available, if required. Although addicts often lapse soon after therapy, they have frequently learned enough to regain self-control, preventing their lapse descending into an uncontrolled relapse. Aftercare or follow-up, in minimal and intensive interventions, usually increases the long-term success rate.

The *Report of an Independent Review of Drug Treatment Services in England* (Task Force to Review Services for Drug Misusers 1996) is cautiously optimistic with respect to treatment outcome, in line with Miller's (1992, 1998) conclusions. Client choice, embracing different options, and client-treatment matching is emphasized in a pluralistic rather than partisan approach. The report offers a useful summary of the various treatment services in England and implicitly emphasizes the systems perspective, where the addict's needs are balanced in relation to the system's needs. It is ethically crucial to bear effectiveness in mind, because patients should be informed of the empirically substantiated advantages and disadvantages of the various options in relation to their personal needs. When pharmacotherapy is used in isolation, patients can fail to develop a sense of personal responsibility and active co-operation in change. If psychological help is offered and medical or social–environmental problems ignored, patients might suffer physical and emotional harm. *Withholding* effective help is, ethically, even less forgivable.

In Britain, the National Treatment Outcome Research Study gives an indication of the sort of effectiveness which current drug treatment can offer (Gossop *et al.* 1999). Both community drug teams and primary health care teams with co-ordinated care were able to engage over four-fifths of heroin users in treatment for at least a month, with outcomes after 6 months of reductions in the use of heroin and other drugs, reduced injecting, reduced drug dealing and improved physical health. The whole community gained. For example, among the patients seen in general practice their involvement in crimes like burglary or street robbery averaged 10.1 days per month at first and crime fell to 2.4 days per month with treatment (a 76 per cent improvement).

Terminating treatment with drug addicts is never simple because there tends to be doubt in the treatment agent and client regarding the future and what has, or has not, been achieved. Similarly, here, the reader should bear in mind the limits involved. This chapter is designed to help the student understand basic elements in addiction treatment. It is not sufficient knowledge, on its own, to treat patients safely and successfully without wider reading, appropriate training and supervision.

A test for your understanding thus far:

A controversial issue in treatment outcome has been the alleged relative importance, or unimportance, of treatment intensity. Professional and non-professional workers alike have tended to emphasize the need for more rather than less help, whilst most researchers question this injunction. Treatment 'intensity' has often been confused with issues of aftercare, a treatment ingredient that isn't a unique feature of intensive help. Miller (1998) highlights the seemingly irreconcilable contrast between those studies finding that patients who stay with treatment for longer tend to do better and those studies using control comparisons where more intensive interventions don't appear to yield better outcomes than less intensive.

- What do you think the rationale is for treatment intensity being linked to outcome?
- Why do you think intensity hasn't always improved results in practice?
- What do you think might be the reasons for the, seemingly discrepant, findings reported by Miller (1998)?

References

Allen, J. P. and Kadden, R. M. (1995) 'Matching clients to alcohol treatments', in R. K. Hester and W. R. Miller (eds) *Handbook of Alcoholism Treatment Approaches: Effective Alternatives*. Boston: Allyn & Bacon.

Álvarez, F. J. and Del Río, M. C. (1999) 'Ultrarapid opiate detoxification: A look at what is happening in Spain', *Addiction*, **94**, 1239–40.

Bandolier (1999) 'Methadone maintenance interventions', *Bandolier*, **6**(8), 7.

Barth, R. P., Pietrzak, J. and Ramler, M. (eds) (1993) *Families Living with Drugs and HIV: Intervention and Treatment Strategies*. New York: Guilford Press.

Bien, T. H., Miller, W. R. and Tonigan, J. S. (1993) 'Brief interventions for alcohol problems: A review', *Addiction*, **88**, 315–36.

Carroll, K. M. (1999) 'Behavioural and cognitive behavioral treatments', in B. S. McCrady and E. E. Epstein (eds) *Addictions: A Comprehensive Guidebook*, pp. 250–67. New York: Oxford University Press.

Commander, M. J., Odell, S. O., Williams, K. J., Sashidharan, S. P. and Surtees, P. G. (1999) 'Pathways to care for alcohol use disorders', *J. Public Health Med.*, **21**, 65–9.

Copello, A. (2000) 'Helping the families of people with alcohol and drug problems', *Primary Health Care*, **10**(2), 23.

Cornish, J. W. and O'Brien, C. P. (2000) 'Pharmacotherapies to prevent relapse: Disulfiram, naltrexone and acamprosate', *Medicine*, **27**(2), 26–8.

Davies, J. B. (1992) *The Myth of Addiction: An Application of the Psychological Theory of Attribution to Illicit Drug Use*. Reading: Harwood Academic.

Department of Health (1991) *Drug Misuse and Dependence: Guidelines on Clinical Management*. London: Her Majesty's Stationery Office.

Department of Health (1999) *Drug Misuse and Dependence: Guidelines on Clinical Management*. London: Her Majesty's Stationery Office.

Egan, G. (1998) *The Skilled Helper: A Problem Management Approach to Helping*. London: Brooks/Cole.

Galanter, M. and Kleber, H. D. (eds) (1994) *Textbook of Substance Abuse Treatment*. Washington, DC: American Psychiatric Press.

Gawin, F. H. and Ellinwood, E. H. (1988) 'Cocaine and other stimulants. Actions, abuse, and treatment', *New Engl. J. Med.*, **318**, 1173–82.

Ghodse, A. H. and Maxwell, D. (eds) (1990) *Drug Abuse Treatment: A Guide for the Caring Professions*. London: Macmillan.

Ghodse, A. H. and McCartney, J. (1992) 'Systems analysis of a drug dependency service in London: St. George's Hospital', *Brit. J. Addict.*, **87**, 1377–85.

Gossop, M., Johns, A. and Green, L. (1986) 'Opiate withdrawal: Inpatient versus outpatient programmes and preferred versus random assignment to treatment', *Brit. Med. J.*, **293**, 103–4.

Gossop, M., Marsden, J., Stewart, D., Lehmann, P. and Strang, J. (1999) 'Methadone treatment practices and outcome for opiate addicts treated in drug clinics and in general practice: results form the national treatment outcome research study', *Brit. J. Gen. Prac.*, **49**, 31–4.

Joe, G. W., Simpson, D. D. and Broome, K. M. (1998) 'Effects of readiness for drug abuse treatment on client retention and assessment of process', *Addiction*, **93**, 1177–90.

Kleber, H. D. (1989) 'Treatment of drug dependence. What works', *Int. Rev. Psychiat.*, **1**, 81–100.

Lindstrom, L. (1992) *Managing Alcoholism. Matching Clients to Treatment.* New York: Oxford University Press.

London, M. and Ghodse, A. H. (1989) 'Types of opiate addiction and notification to the Home Office', *Brit. J. Psychiat.*, **154**, 835–8.

McCartney, J. (1996) 'A community study of natural change across the addictions', *Addict. Res.*, **4**, 65–83.

McCartney, J. (1997) 'Between knowledge and desire. Perceptions of decision-making and addiction', *Substance Use and Abuse*, **32**, 2061–93.

McVinney, L. D. (ed.) (1997) *Chemical Dependency Treatment: Innovative Group Approaches.* New York: Haworth Press.

Miller, W. R. (1992) 'The effectiveness of treatment for substance abuse: reasons for optimism', *J. Substance Abuse Treat.*, **9**, 93–102.

Miller, W. R. (1998) Why do people change addictive behaviour? The 1996 H. David Archibald Lecture. *Addiction*, **93**, 163–72.

Miller, W. R. and Rollnick, S. (1991) *Motivational Interviewing. Preparing People to Change Addictive Behaviour.* New York: Guilford Press.

Moos, R. A., Finney, J. W. and Cronkite, R. C. (1990) *Alcoholism Treatment. Context, Process and Outcome.* Oxford: Oxford University Press.

Morgan, J. R. (1990) 'Clinical management', in A. H. Ghodse, and D. Maxwell (eds) *Drug Abuse Treatment: A Guide for the Caring Professions.* London: Macmillan.

O'Farrell, T. J. and Fals-Stewart, W. (1999) 'Treatment models and methods: Family models', in B. S. McCrady and E. E. Epstein (eds) *Addictions: A Comprehensive Guidebook*, pp. 287–305. New York: Oxford University Press.

Oppenheimer, E., Sheehan, M. and Taylor, C. (1988) 'Letting the clients speak: Drug misusers and the process of help seeking', *Brit. J. Addict.*, **83**, 635–47.

Prochaska, J. O. and Di Clemente, C. C. (1982) 'Transtheoretical therapy: Toward a more integrative model of change', *Psychother. Theory, Res. Prac.*, **19**, 276–88.

Schneider, R. J. and Khantzian, E. J. (1992) 'Psychotherapy and patient needs in the treatment of alcohol and cocaine abuse', in M. Galanter (ed.) *Recent Developments in Alcoholism*, Vol. 10. New York: Plenum Press.

Scott, M. (1989) 'Relapse prevention training', in G. Bennett (ed.) *Treating Drug Abusers.* London: Routledge.

Task Force to Review Services for Drug Misusers (1996) *Report of an Independent Review of Drug Treatment Services in England.* Wetherby: Department of Health.

Vaillant, G. E. (1988) 'What can long-term follow-up teach us about relapse and prevention of relapse in addiction', *Brit. J. Addict.*, **83**, 1147–57.

Wanigaratne, S., Wallace, W., Pullin, J., Keaney, F. and Farmer, R. (1990) *Relapse Prevention for Addictive Behaviours: A Manual for Therapists.* London: Blackwell Scientific.

West, S. L. (2000) 'Review: Studies with sufficient follow up do not show a clear benefit for pharmacotherapy in alcohol dependence', *Evidence-Based Mental Health*, **3**, 15.

White, M. (1998) 'London borough's drug treatment support a success', *Community Care*, 1 October 1998, 7.

Chapter 17

Rehabilitation: the long haul

John Chacksfield

Give a man a fish and you feed him for a day. Teach a man to fish and
you feed him for a lifetime.

(Ancient Chinese saying)

Key points

*Rehabilitation in the field of addictions is an intensive process that involves
a return to balanced living following dependent or problem use of legal and
illegal drugs or certain repetitive behaviours, such as gambling. It involves
the post-treatment development of the ability to live independent from the
preoccupation with, salience of or dominance by an addictive behaviour.
The following key processes are involved:*

- *A multidisciplinary and multi-modal approach.*
- *A comprehensive assessment and rehabilitation plan.*
- *Education and health promotion.*
- *Gradual and supported change in attitudes, behaviour and skills over
  time.*
- *Assistance to build support systems and coping strategies for the main-
  tenance of independent living.*

Conceptualizing addictions rehabilitation

Rehabilitation can be described as returning to life after an illness or disability. This and the associated concept of '*habilitation*' (which deals with
the new) are concepts central to a professional approach to enabling
recovery from addictions. In the field of addictions, rehabilitation is generally client focused. A useful definition of *client-centred rehabilitation* has
been offered by McColl *et al.* (1997) as:

A therapeutic orientation whereby clients engage the assistance and support of a therapist to facilitate their problem solving and the achievement of their own goals.

The importance of taking a client focus in addictions rehabilitation lies in the need for an addicted person to take responsibility for engagement in their treatment. Those who are able to play their part in recovery, rather than simply relying on a treatment agency 'to do the work', generally discover ongoing success rather than repeated disappointment and failure in recovery.

This need for the client to take responsibility for, and ownership of their part in, treatment is central to most approaches to addictions rehabilitation and will be referred to throughout the chapter.

The rehabilitation process

The task of maintaining abstinence from an addictive behaviour (or of establishing a strictly controlled regime), with its frequent failures, is what makes rehabilitation a lengthy process. It is the frequency of these failures that can bring despair to both client and health professional, yet, when success occurs in the face of such a major challenge, it is invariably long-lasting and highly satisfactory.

The business of rehabilitation invariably leads to personal development in both *therapist and client*. It is this ground for growth that makes addiction work so rewarding and interesting. The secret of successful rehabilitation lies in tackling the subtle problems of addiction with the effective combination of a comprehensive variety of interventions, tailored to meet the individual needs of the client and using their own effort. Edwards (1987) advises:

Therapeutic work is only likely successfully to produce movement when its efforts are in alignment with the real possibilities for change within the individual, his family, and social setting. ... The basic work of therapy is largely concerned with nudging and supporting the movement along these 'natural' pathways of recovery. ... the clumsy therapist is like someone who tries to carve a piece of wood without respect for the grain.

Therapeutic tools and facilitating change

Most rehabilitation is about helping the client to initiate and adjust to change. Change often needs to occur in factors like attitude, lifestyle, habits and roles (Rotert 1990), all of which have developed alongside an addiction and have grown to support it. The national strategy *Tackling Drugs to Build a Better Britain* prioritizes this need to:

Support problem drug misusers in reviewing and changing their behaviour towards more positive lifestyles – linking up where appropriate with accommodation, education and employment services.

(Central Drugs Coordination Unit 1998)

Usually, for most people, addressing change is anxiety provoking, especially where it will affect long-established habits on which someone has based their sense of security. It is clear that therapists must develop a good rapport with a client if they are to discover what the real possibilities for change are within an individual as well as to understand that individual's life context (family and social setting).

Concepts which address change

This focus on change in addictions rehabilitation has formed the basis for four key concepts which have been developed and successfully used within the field to facilitate graded progress:

- *The Minnesota Model* (described by Royal College of Psychiatrists 1987), which focuses on intensive treatment and emphasizes the disease concept of addiction, abstinence and the Twelve Steps programme of Alcoholics Anonymous (AA) or Narcotics Anonymous (NA).
- *Relapse Prevention* (Marlatt and Gordon 1985; Gossop 1989), which focuses on teaching the client how to anticipate and cope with the pressures and problems that may lead towards a relapse.
- *Motivational Interviewing* (Miller 1983; Miller and Rollnick 1991), which focuses on encouraging motivation for change through a specific counselling technique based on the Model of Change.
- *The Model of Change* (Prochaska and Di Clemente 1986, 1983), an approach which describes the stages through which a person passes when changing addictive behaviour and can be used as a structure for observing, tracking and facilitating rehabilitation.

These concepts, which will be explained later, have formalized what addictions therapists and fellow workers have developed through years of convoluted experience. More recently, many other less highly profiled, although equally effective, 'brief' counselling approaches have also developed along these lines: such as Solution Focused Therapy (Berg and Miller 1996), Rational Emotive Therapy (Ellis *et al.* 1988), Cognitive Analytic Therapy (Ryle 1991) and Perceptual Adjustment Therapy (Holder and Williams 1995). Other approaches, such as Cue Exposure Treatment (Drummond *et al.* 1995), can form an adjunct to a relapse prevention programme.

Focus on empowerment

All the approaches above are centred on enabling the individual, who has the addiction, with the skills to maintain complete abstinence or a balanced pattern of behaviour. All these approaches are usually applied via the media of groupwork and individual counselling. Some treatment agencies, such as Turning Point, have maximized the impact of these philosophies by coupling them with Maxwell Jones' concept of Therapeutic Communities (Rosenthal 1991).

The availability of a range of approaches is important in both treatment and maintenance of the results (Edwards 1987). The wide range of needs presented by someone with an addiction needs a multifaceted approach which meets such varied needs. Failure to offer multidisciplinary and multi-modal treatment can lead to patients being forced to fit the treatment being offered (Gossop 1994).

Three aspects of rehabilitation

Rehabilitation is divided into three main areas of concern:

- addiction rehabilitation goals;
- approaches to the rehabilitation process;
- maintenance of life balance.

In each of these sections, theory and practice will be examined followed by some specific issues which may give cause for tailoring treatment in specific ways (e.g. age, subcultural and transcultural considerations or dual diagnosis).

Addiction rehabilitation goals

Max Glatt, in his *Tentative Chart of Addiction and Recovery* (1974) (see Figure 17.1), suggests that, following a brief period of detoxification and treatment, *recovery takes a good 3 to 8 years*. This period necessitates extensive work to change both personal developmental or *intrinsic factors*, such as the ability to cope with triggers relapse and craving, and *extrinsic factors*, including repair of the damaging impact of addiction on the addict's family and immediate society. These factors form the basis for rehabilitation goals that are addressed in a wide variety of ways.

Approaches to planning rehabilitation

Following a period of detoxification and orientation to the rehabilitation process, most addiction treatment agencies will aim for a similar set of goals,

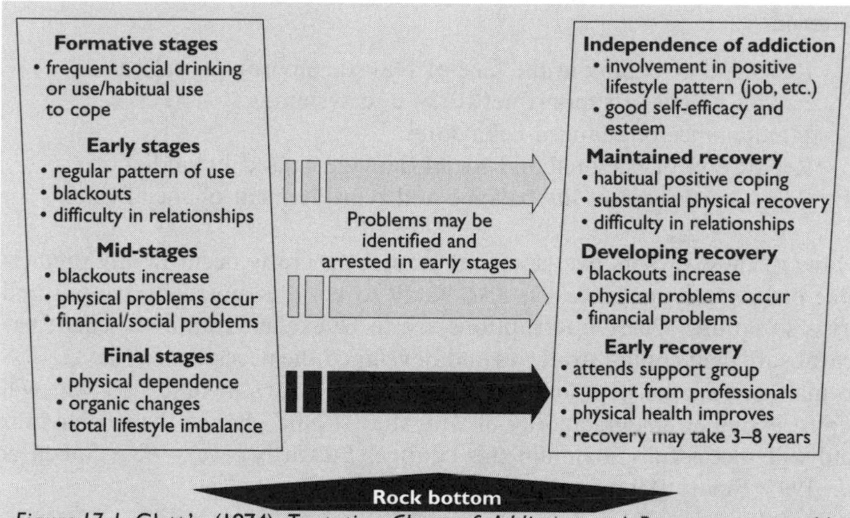

Figure 17.1 Glatt's (1974) *Tentative Chart of Addiction and Recovery*; a possible representation of the long haul from addiction to recovery (after M. Glatt's (1974) *Tentative Chart of Alcohol Addiction and Recovery*; reprinted with permission from Hodder & Stoughton).

modified in application to suit the particular set of problems presented by a client. Most rehabilitation will start with helping an individual understand that they are addicted. This is then followed by the identification of specific goals for recovery.

Key focal areas

A number of key focal areas for rehabilitation can be identified as follows:

Internal

- Recognition of the problem.
- Preparation to change – establishment of reasons and potential risk situations.
- Development of strong internal motivation to create and maintain independence from addictive substances and behaviours.
- Development and rehearsal of skills for coping with intrinsic cues (thoughts and feelings).
- Identification and practice of coping strategies for risk of relapse.
- Building self-esteem and personal confidence.
- Building and development of life skills and habits in leisure, self-maintenance and work.

External

- Rehearsal of coping in the face of realistic environmental cues.
- Establishment of support networks and systems.
- Maintenance of changed behaviour.
- Reparation of personal and social damage caused by addiction.
- Re-establishment of life balance and reinstatement of social roles.

Many of these factors overlap. They do not generally occur in any predictable order, and the process is also likely to be interrupted by relapse and crisis situations. These interruptions are to be expected until the client has learnt sufficient coping strategies and developed the necessary self-awareness to maintain a consistent level of changed behaviour. At times, a client will report a sudden understanding of why they should change their behaviour and will successfully maintain this change (Edwards et al., 1987; Sobell et al., 1993; Smart 1994).

The beginning: recognition of the problem

Of paramount importance is the need for the client to begin to take some responsibility for engaging in rehabilitation.

A key assumption about the Dependence Syndrome (Edwards and Gross 1976) is that it is incurable and can only be arrested. As stated by the Royal College of Psychiatrists about alcohol dependence (1986), *once a person has become severely dependent there is often little choice between continuing to drink destructively or stopping drinking altogether.* This statement would probably be applicable to any addictive behaviour and total abstinence is often the only realistic treatment goal. The idea of abstinence is, however, one of the most difficult concepts for the client to accept.

Acceptance and denial

A maxim of both AA and NA is that the addict must accept the fact that they are an addict; for example, people with alcohol dependence attending an AA meeting open a contribution to the discussion by introducing their name, immediately followed by the words 'I am an alcoholic'. It could be argued that the continual reinforcement of this statement never frees the person from a disease focus. However, proponents of AA would say that until a person has accepted the fact they are addicted to alcohol, indicated by a willingness to say 'I am an alcoholic', they will never begin to recover. This first step to recognition of responsibility also forms the first step of any Twelve Step programme (see Table 17.1). The seemingly 'quasi-religious' language used by Twelve Step groups, such as AA, may repulse some health professionals new to the addictions field, who are used to a more

Table 17.1 The Twelve Steps

1	We admitted we were powerless over alcohol/drugs (or another addiction) – that our lives had become unmanageable.
2	Came to believe that a Power greater than ourselves could restore us to sanity.
3	Made a decision to turn our will and our lives over to the care of God as we understood Him.
4	Made a searching and fearless moral inventory of ourselves.
5	Admitted to God, to ourselves and to another human being the exact nature of our wrongs.
6	Were entirely ready to have God remove all these defects of character.
7	Humbly asked Him to remove our shortcomings.
8	Made a list of all persons we had harmed, and became willing to make amends to them all.
9	Made direct amends to such people wherever possible, except when to do so would injure them or others.
10	Continued to take personal inventory and when we were wrong promptly admitted it.
11	Sought through prayer and meditation to improve our conscious contact with God as we understood Him, praying only for knowledge of His will for us and the power to carry that out.
12	Having had a spiritual awakening as the result of these steps, we tried to carry this message to alcoholics and to practise these principles in all our affairs.

scientific presentation of treatment media. It is important that such reactions are evaluated and transgressed allowing a clear understanding of underlying principles to enhance and inform the professional's own treatment repertoire.

The reality is that the Twelve Step approach can be highly effective as a graded sequence of rehabilitation goals which lead to life reorganization and increased self-efficacy and independence. Furthermore, these approaches involve support from an extensive network of non-substance-using addicts, whose personal experience of addictions is, perhaps rightly, often considered more helpful than the training of the health professional. It is advised that health professionals, working with such complex problems as those related to addictions, learn to make use of any approach which works for a particular client, or as Ryle, father of Cognitive Analytic Therapy says, to 'push where it moves' (Ryle 1991).

Denial, avoidance and other ego-defence mechanisms

A further reinforcement of the need for recognition of the presence of an addiction (and related problems) by the person who is alcohol or substance dependent is the recognition of ego-defence phenomena. Freud invented the concept of an 'ego-defence mechanism'. Ego-defence mechanisms (EDMs) such as denial, avoidance or displacement are part of the make-up of every human psyche. They serve to protect the ego from psychological stress, and can be a useful part of adaptation to conflict, regulating and facilitating emotional maturation (see Figure 17.2, p. 242); for instance, a newly qualified health professional, when presented with their first alcoholic client, may find themselves feeling 'defensive' or wishing to avoid the client for no apparent reason. It is important that these reactions are faced and discussed within supervision sessions or staff support groups (Brown and Brooke 1991; Velleman 1992). These reactions are normal; indeed, the more the professional learns about his or her own psyche, the better equipped they will be to assist the client.

Where an individual client's defence mechanisms have become overwhelmingly adapted to protect the ego from the painful recognition that they have an addictive behaviour, the EDM system has clearly become pathological. This phenomenon (an example of *denial*) is observable in most addicts and must be addressed if any successful rehabilitation is to take place. Here is Mary, a 40-year-old woman who is known to drink two to three bottles of whisky a day. She is denying, displacing and avoiding her feelings about the problems arising from her alcohol dependence:

> Mary has been referred to an alcohol outpatient clinic. She is angry towards her social worker for suggesting her alcohol use is causing her problems looking after her children. 'You professionals keep telling me to stop the drink, but how else am I supposed to cope with my boy? He's the problem! He's always playing truant and trying to run away to his Aunt's!'

According to Kearney (1996), denial is probably the defence which forms the first major barrier to self-recognition and further healing of the intrinsic and extrinsic effects of addiction. One of the most effective methods for countering denial, and other EDMs, is through group therapy (Levine and Gallogy 1985). The addict who meets other addicts in a similar predicament to her or his own is far more likely to begin to recognize and comprehend that predicament than if simply told about it by a health professional. It is significant, however, that once the first glimmerings of recognition of the problem occur, rehabilitation can begin in earnest.

Approaches to the rehabilitation process

When rehabilitation begins

The process of rehabilitation *per se* can begin from a variety of points – before, during or after detoxification. Although a clinical team may demark the point at which, in their view, they have stopped detoxification and 'treatment' and rehabilitation has started, in reality true rehabilitation only begins when the client is ready. This can have occurred before they even began detoxification, where they have made the decision to control their addiction, have begun to work out how they will support change and are ready to seek professional help. This decision can also occur at other stages.

The primary principle of addictions rehabilitation, therefore, is probably that rehabilitation must begin with the client. Focus on the client is important in three respects:

- *Responsibility* – the client has ultimate responsibility for their rehabilitation;
- *Pace* – rehabilitation must progress at the client's pace;
- *Situation* – Rehabilitation must be aligned with the client's life situation and its impact on their volition.

These three factors are essential if a successful rehabilitation process is to be initiated, and they depend on a comprehensive assessment.

A global approach to rehabilitation

A useful global approach to assessing, planning and evaluating rehabilitation has been developed by occupational therapists and is based on systems theory. Known as the Model of Human Occupation (Kielhofner 1985; 1995), its basic premise is to view the individual as an 'open system' (see Figure 17.2), comprising internal volition, habituation and performance systems, that interacts with their social and physical *environment*. For people dependent on alcohol, it has been shown that their social environment is a crucial factor in determining their quality of life (Foster *et al.* 1999).

Leckie (1990) discusses the usefulness of systems thinking in working with problem drinkers. He says systems thinking helps the professional to view the drinker within the context of a variety of overlapping and interrelating social systems. The same idea will be true of other client groups.

The Human Open System

Although the entire Model of Human Occupation is complex, the concept of the Human Open System (Figure 17.2) can be easily adapted to suit the assessment and treatment of addictions (Rotert 1990). It is sufficiently adaptable to allow the integration of other addictions approaches, such as Relapse Prevention and the Model of Change. This model views the human being as an open system and describes how this person (system) interacts with the social and physical aspects of the surrounding environment. The

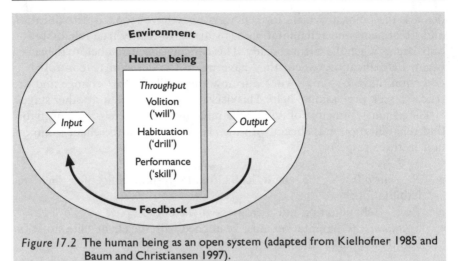

Figure 17.2 The human being as an open system (adapted from Kielhofner 1985 and Baum and Christiansen 1997).

way an individual interacts with their environment can occur within three broad dimensions or domains of behaviour as follows:

- *work or productivity* – including any goal-oriented behaviour, often done out of necessity, such as work for monetary reward, voluntary work or housework;
- *self-maintenance* – including self-care, domestic skills and personal financial management;
- *leisure* – including any activity done out of personal choice and usually for its intrinsic benefits (i.e. for 'fun').

Assessment of the nature and range of the three subsystems affecting *throughput* and their relationship to the environment in each individual case forms the basis for a comprehensive, client-centred *rehabilitation plan*.

The Kielhofner (1985) system also works well as a structure for different tools for assessment, whether as a team approach or by an individual therapist. It is important that, during these formative stages, the therapist both assists the client to be aware of the presence of the problem and gives the client a realistic view of rehabilitation; for example, that it can take a long time but, with effort, the process can be shortened.

Other models

This and other global rehabilitation models are in use within the addictions field. They are useful especially for developing the professional's thinking in planning and organizing intervention. It is important to recognize, however, that to rely on one specific model is to severely limit the range of skills available for rehabilitation, whereas an eclectic approach will best serve the variety of needs presented by the client.

Other rehabilitation models that have been applied in the addictions field include the Canadian Model of Occupational Performance (CAOT 1993) and The Human Needs Model of Nursing (Minshull *et al.* 1986).

Assessment for rehabilitation planning

Initial considerations

Often, a screening assessment will be carried out, consisting of a loosely structured interview. This will then be followed by a fuller, more structured multidisciplinary assessment. A detailed assessment may take several hours of interview and discussion sessions while a range of assessment methods are implemented, followed by observation in group activities, such as task-oriented groups involving practical skills, or other treatment settings. In practice, different members of a treatment team will often carry out different aspects of the assessment. It is important that the rehabilitation plan is co-ordinated by an experienced member of the team, regardless of profession or position. This idea is becoming increasingly evident within community addictions teams, in part due to staff shortages and partly due to overlap in professional remit (see 'Professional input – finding the balance' on p. 252).

Of further importance is the role of a key worker as a named co-ordinator for the client. However, this has to be balanced with the need for the client to be encouraged to identify with the addictions *team* as their contact point rather than a specific individual. This is important in case of the following:

1   staff changes;
2   avoidance of responsibility by a client through the fostering of dependence on an individual staff member.

The latter point needs to be discussed frequently within staff supervision.

Approaches to screening assessments

This phase should take place after any initial detoxification and physical 'overhaul'. A basic initial interview is carried out, followed by a brief focus on specific techniques and strategies. The team will need to keep in mind the following:

• what the client sees as the key issues;
• risk situations which may precipitate harm and relapse;
• extent of motivation for engaging in the rehabilitation process;
• past history of rehabilitation attempts;

- the potential for return from addiction living to balanced living;
- the individual strengths and weaknesses which can influence progress;
- ability to interact with other people;
- available and needed support and achievement of community integration;
- client's capacity for establishment of 'warning systems' for coping with relapse.

Once an initial list of needs and problems has been drawn up, this can be discussed by the rehabilitation team, who can then decide on the readiness of the client for the next stage of rehabilitation. Following this, a fuller assessment is carried out with the goal of formulating a comprehensive rehabilitation plan.

Broadening the assessment

The full assessment generally aims to extend the exploration of key factors highlighted within the screening interview. This assessment often takes the form of in-depth interviews and use of questionnaires.

At first, it is important to establish the extent of the client's motivation (*volition*). In Motivational Interviewing, the client is assisted to draw up a personal inventory of problems associated with the abuse. This quickly clarifies important issues for both the patient and the therapist and leads to more productive work (Royal College of Psychiatrists 1987). Motivational Interviewing, to be effective, requires extensive training and practice. A quicker, simpler, initial assessment tool is the Relapse Prevention Matrix (RPM: Marlatt and Gordon 1985). The RPM is usable by both therapist and client and consists of a grid of two columns and two rows (see Table 17.2). As a decision matrix, it can be a useful starting point to a therapeutic relationship.

Another tool that can assist therapist and patient to structure an 'overview' of the problems they face is The Stages of Change Model (Prochaska and Di Clemente 1986). This tool is useful throughout the rehabilitation

Table 17.2 Example of a Relapse Prevention Matrix (based on ideas by Marlatt and Gordon 1985)

	Positive	Negative
Reasons for continuing to use cocaine		
Reasons for stopping use of cocaine		

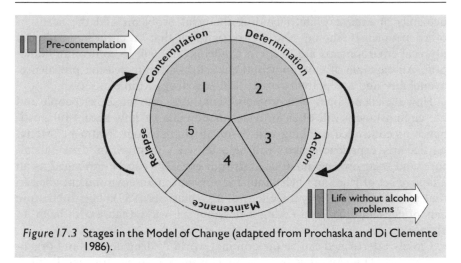

Figure 17.3 Stages in the Model of Change (adapted from Prochaska and Di Clemente 1986).

process, and frequent reference to it in counselling sessions (Velleman 1992) can be useful in assisting the client to develop insight into their situation and for the therapist to keep track of progress. The Stages of Change model identifies five key points in the process of change from blind dependence on an addictive substance to abstinence (see Figure 17.3). These range from *precontemplation, contemplation, determination, action, maintenance and relapse* and are usually represented as a cycle.

Exploration of issues surrounding change, its maintenance and the obstacles the client sees as preventing it can usefully lead to assessment of the *habituation* and *performance* subsystems in the Model of Human Occupation (Kielhofner 1985). These concern the behaviour that the client has learned to use to interact with the environment, resulting from their personality traits and life experiences. Assessment of these aspects involves drawing on both recent and past history. Three particularly important areas interrelate with each other and should be considered as important focal areas for rehabilitation:

- *coping* – life skills, skills for coping with sources of stress and support systems;
- *reinforcement* – sources of motivation and meaning in life other than the drug;
- *environment* – the life environment including productivity, leisure and self-maintenance aspects.

Coping, reinforcement and environmental influence

Probably, the first part of the client's 'life picture' seen by the professional is that of the *disability* caused by an increasing failure to cope with the demands of everyday life, including financial problems and the needs of other people. It should also be recognized that a client's social and physical environments are likely to challenge any will to control an addiction. An example of environmental challenge is the increasing prevalence, availability and acceptability of alcohol and drugs within society.

How the client copes with problems, emotions, craving, other people and life circumstances will often provide an estimate of how successful abstinence or controlled drinking will be on discharge from treatment. Many, particularly experienced users will be extensively involved in 'drug subculture' and their own community of drug users. This can be explained as an adaptive set of rules and behaviour for survival in an environment alien to that of the clinician (Royal College of Psychiatrists 1987). This subculture can often reinforce drug-taking behaviour in ways that are difficult to change.

Closely intertwined can be the criminal world of drug dealing and organized supply. This can range from those who shoplift to pay for their habit to serious involvement in drug-dealing gangs and violent crime. The latter cases are rarer in most treatment agencies. The issue of the relationship between drug use and crime has been extensively discussed (D'Orban 1986; Bennett et al. 1996; Taylor and Bennett 1999; Goldstein 1986). As well as reinforcing addictive behaviour, involvement in crime, whether on the borderline or more deeply, can form a type of pseudo-social support for many drug users.

Family considerations

Closer to home, many families of addicts will have developed mechanisms for coping with the addicted person's behaviour, such as the long-suffering partner who puts up with his wife's rages and drunken behaviour, simply for a 'quiet life'. This unproductive type of coping becomes a habit and is termed *co-dependence* (Dittrich 1993).

An alcohol-dependent person who returns after treatment to a family which has developed co-dependence will often meet with what he feels are bizarre reactions. These reactions occur simply because the family is unused to the very different behaviour of the 'dry alcoholic'. Unfortunately, the bizarre behaviour of the co-dependent family can include cues which may trigger a return to drinking behaviour. Here we find the client who relapses after a row where his partner does not expect him to do the dishes. His normal reaction to these rows has always been to go and 'drown his sorrows' in alcohol. He reacts to the cue the row presents without thinking and, before he knows it, he is back in the pub. Other family and relationship problems often occur in relation to intimacy and sexuality (Cutland 1990; Hetherington 1995). Discussion of these issues forms an important aspect of assessment. Furthermore, if a client gives permission, it can be very useful to obtain historical information about them from family members.

A number of authors discuss family therapy and the assessment and management of family issues in depth (e.g. O'Farrell 1993; Steinglass *et al.* 1987; Kaufman 1985). The inclusion of family aspects is important as it allows the rehabilitation planner to see the client *in context*. It will be important to explore other social support mechanisms, including friends, workplace support mechanisms and prior use of self-help groups or treatment agencies.

Others may have very minimal or no support, such as the secretive home drinker or the solitary, unemployed user who sees drugs as the only way of dealing with an ongoing meaningless existence. Although the responsibility for relapse to addiction must be placed on the client, it should be remembered that the treatment team has a role to play in establishing and assisting the client to develop support systems. The client will also need help reorganizing support systems which may no longer be effective, such as the family who is not used to non-drinking behaviour.

Coping and reinforcement of addictive behaviour

Drug-using thinking and behaviour can be seen to be reinforced in two main ways: externally and internally. The role of reinforcement in addiction has been examined extensively (Drummond *et al.* 1995). It is important to consider reinforcement issues, as so many addictive behaviours are associated with everyday social events and internal moods and thoughts. These social events can become cues, or triggers, to addictive behaviour and present considerable problems in social rehabilitation.

External reinforcement

External reinforcement has been examined extensively within cue exposure theory (Drummond *et al.* 1995; Siegal 1999). Here we find the alcoholic who enters a public house ostensibly to buy a packet of cigarettes and ends up drinking with his friends. Analysis of this seemingly random event will show that specific environmental cues, such as the sight and smell of alcoholic drinks, the social 'atmosphere' or the sight of a particular person, have initiated the actions that led to consumption of alcohol.

Internal reinforcement

It is highly likely that, intertwined with these environmental cues, are *intrinsic* cues such as mood state (e.g. celebratory, angry, depressed, guilty, anxious), particular thoughts and memories or physiological state (e.g. feeling hungry or tired). These factors will further contribute to the chain reaction, ending in a drink. Cue Exposure Treatment (Drummond *et al.*,

1995), a method originally widely used for treating phobias, aims to assist the client to reduce their reactions to internal and external stimuli which might precipitate a relapse.

These concepts are extensively discussed within the literature on Relapse Prevention (Gossop 1989). Closely linked to the concepts of intrinsic and extrinsic reinforcement are the concepts of internal and external locus of control.

Locus of control

The concept of Locus of Control, the full title being Locus of Control of Reinforcement, was initially developed by Rotter (1966). A person can be seen to have *internal* or *external* locus of control; *internal* being when someone attributes responsibility for everyday happenings and events (including crises) to themselves. This can be positive in terms of clear acceptance of responsibility or negative in terms of a fatalistic attitude and failure to act to change behaviour. *External* locus of control is the opposite and often involves blame of other people or circumstances for the afflictions of the drug or alcohol user. Closely linked with denial, this latter state of mind can mean the person also believes failure of treatment to be purely caused by the treatment agency rather than their own failure to follow rehabilitation goals.

The issues surrounding external locus of control further emphasize the importance of client-centred treatment and the client's understanding of their own responsibility during rehabilitation.

Self-concept

In tandem with locus of control will be self-awareness and insight, self-esteem, self-efficacy and life meaning. These concepts generally refer to a person's view of themselves within the world. Carl Jung says '*Man cannot stand a meaningless life*' (Fordham 1966). As well as simple lack of insight, it is likely that a person who derives life meaning from habitual gambling, pathologically obsessive patterns of sexual activity, use of alcohol or drugs will not give up their habit easily.

Structuring assessments for effective rehabilitation planning

The use of a variety of assessment techniques, in conjunction with a structure such as the Model of Human Occupation (Kielhofner 1985), can provide a comprehensive base for rehabilitation planning. The plan will then focus on assisting the client to become as independent as possible from drug or alcohol use and either develop this or aim to maintain them at their current stage of rehabilitation. Certain key issues will be commonly identified during assessment. Some of these are discussed below.

The importance of social history

Considering the significance of external triggers to drug-using behaviour and external reinforcers, it will be highly important to examine the client's social history in context, or 'as they tell it', and what has been called the addictions 'career' (Raistrick 1991). Here is to be found the value of the unstructured or semi-structured assessment interview.

A flavour of the emotions attached to circumstances – the childhood situation, their successes and failures, whom a client experienced problems with and how they survived – is all part of this information. Skinner (1978) describes these conditions as 'contingencies of reinforcement', because they led to the development of specific behaviour patterns.

The behaviour patterns identified in the social background will be highly significant in treatment outcome, reinforcing or nullifying the impact of treatment (Moos *et al.* 1979). Knowledge of these established behaviour patterns (*habituation subsystem*) and skills developed for interacting with the environment (*performance subsystem*) can allow them to be usefully shaped towards rehabilitation. To augment the unstructured assessment of the social background, a number of structured questionnaires are useful. Examples of these are shown in Table 17.3. A wide range of life skills and aspects of everyday interactions with people and environment will be affected by misuse of or dependence on a drug. Assessment by questionnaire can only ever form part of the picture. Extensive assessment through interview *and observation*, individually and within group activities, is therefore essential.

Table 17.3 Examples of useful structured questionnaires

Title	Author(s)	Available from
The Alcohol Problems Questionnaire	Drummond 1991	National Addiction Centre
The Coping Responses Inventory	Moos 1993	NFER Nelson
Drinking-Related Locus of Control Scale (DRIE)	Keyson and Janda (see Kivlahan et al. 1983)	
Substance Abuse Assessment Questionnaire	Ghodse 1989	Department of Addictive Behaviours, St George's Hospital Medical School
The South Oaks Gambling Screen	Lesieur and Blume 1988	Reproduced in Glass (1991)
The Family Environment Scale	Moos et al. 1979	

Special issues

Special issues for consideration include:

- gender-related phenomena;
- physical illness and disability;
- age-related aspects;
- cultural considerations;
- dual diagnosis of mental illness;
- poly-substance use.

Almost all of these factors can influence the way an individual perceives their addictive behaviour, such as in terms of its harmfulness or their motivation to change. Common examples include the greater effect of alcohol on women due to their size and biochemical make-up (Kent 1990). Another example is the older person who views drinking spirits as more dangerous than beer or wine. A teenager may see designer drugs, such as Ecstasy, as a ticket to social acceptability and confidence or simply to feel better about themselves (Henderson 1997). A Rastafarian might use cannabis as part of his religious reflection, or there is the recent example of the person with multiple sclerosis who gains symptomatic benefit from smoking a joint.

A good example of using client-centred rehabilitation skills to improve the quality of life of patients who raise special issues has been reported by Willenbring (2000). The skills of goal setting, working in different environments, problem solving and the involvement of families helped American men with a severe physical illness related to alcohol who were still drinking heavily before this 'integrated' intervention.

Furthermore, the different 'culture' (including language, style of dress, behaviour patterns) of drug- and alcohol-using groups may be extremely different to that of the health professional. Understanding the language of 'drug culture', for instance, can assist the therapist to establish rapport and engage the client in significant ways (Davies 1997). A client may talk about 'works' (equipment for injecting heroin) or 'snorting whizz' (inhaling amphetamines), for instance. Being able to understand these terms may have started a discussion around the potential harm surrounding injecting, which may never have happened had the therapist not been able to 'tune in' to what their client was saying.

Dual diagnosis of mental illness and substance-use disorders

Dual diagnosis presents a special issue for consideration. Several studies have shown that people with an addiction who also have schizophrenia or another mental illness are twice as likely to relapse and require readmission to hospital (Caan and Crowe 1994). Furthermore, the Epidemiological Catchment Area Survey (Swanson *et al.* 1990), which surveyed more than 1,000 Americans with a mental illness, demonstrated that addition of addictive behaviour to mental illness can increase the risk of violence to people and property threefold. Many patients with co-morbid mental illness and substance problems are admitted to prison over and over again (Birmingham 1999).

Within some groups of psychiatric patients, dual diagnosis is surprisingly common; for example, 61 per cent of Americans with bipolar disorder develop this 'double trouble' (Rosenthal and Westreich 1999). Dual diagnosis also presents motivational issues. Often, people with a mental illness are known to use alcohol and other drugs to comfort the distress and other effects caused by the symptoms of their illness; for example, the Royal College of Psychiatrists has identified a population of men who escalate their drinking during periods of depression as 'self-medication'. Here, addictive behaviour can lead to a vicious cycle where prescribed medication is taken erratically or not taken at all, depressive symptoms worsen, addictive behaviour increases and a major crisis and readmission to hospital is precipitated. Recent research (e.g. Driessen *et al.* 1998; Appleby *et al.* 2001) has linked alcoholism to increased risk of suicide when associated with mental illness.

Fortunately, dual diagnosis is no longer the label of despair which it used to be, and models of good practice are emerging. Gafoor and Rassool (1998) give a clear framework for assessment by a mental health nurse, incorporating an appropriate mental state and physical examination. About a third of Holland's (1998) patients with schizophrenia also had drink and drug problems. A new partnership between his community mental health team and local addiction services led to joint working, including structured daily living programmes and befriending schemes to rehabilitate this socially excluded group. The homeless are especially vulnerable to co-morbid problems and the Addiction Recovery Project in Camden has developed 'floating support workers' to help people, who have just finished detoxification, with their housing, benefits and even mediation with neighbours who object to their presence. One of the most impressive *integrated* interventions was evaluated by McLellan *et al.* (1998) in Philadephia. Drug group counsellors worked together with social work case managers for individual clients and there were improvements in the severity of addiction, mental health and physical health problems.

Rehabilitation – the toolkit

Rehabilitation is a 'long haul' primarily as it has to account for multiple factors affecting the client on discharge, and this takes time. It may be impossible to meet every need; however, the importance of documenting as much detail as possible within the clinical picture is highly useful for future professionals who may encounter the same client. Also, what is a low-priority problem now may be high priority later when a client's circumstance or view of their addiction changes.

Multidisciplinary notes

The use of multidisciplinary clinical notes can be particularly useful for a professional who wishes to quickly view the extent of involvement from each member of the team. This kind of clinical note also makes team-generated lists of treatment aims easier to maintain and re-evaluate.

The multidisciplinary team

Clearly, time constraints will prevent any single professional from completing a full assessment of rehabilitation needs for all of his or her clients, just as one professional would be unable to offer all aspects of treatment. The value of multidisciplinary and multimodal approaches is paramount in this.

Professional input – finding the balance

The tools and techniques at the disposal of the multidisciplinary team can be described as falling into two categories, namely *addiction-specific* interventions and *profession-specific* interventions (see Figure 17.4). Successful planning of rehabilitation occurs where there is a meeting of the two categories which integrate, to a greater or lesser extent, according to the individual needs of the client and the rapport they may have with a specific professional.

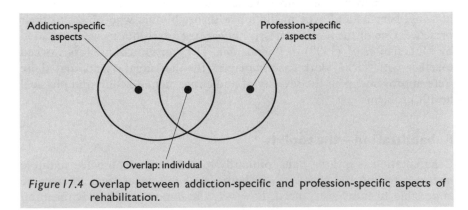

Addiction-specific
aspects

Profession-specific
aspects

Overlap: individual

Figure 17.4 Overlap between addiction-specific and profession-specific aspects of rehabilitation.

Addiction-specific interventions

Techniques such as relapse prevention, motivational interviewing and harm minimization fall into the 'addiction-specific' category. These techniques can be practised by any professional trained in them, although some may feel their professional training to be more compatible with certain addiction-specific interventions than others.

Profession-specific intervention

Profession-specific interventions tend to relate to the unique skills each separate professional in the team can contribute, based on their own training. A good example might be the pharmacological and prescribing knowledge unique to the medical profession or the knowledge of social and family factors which a social worker might offer.

Voluntary sector support

As well as trained professionals, the involvement of personnel from the voluntary sector and family members in discharge planning generally contributes to the successful support of a client on return to the community. Galanter (1987) discusses research that shows large self-help groups, such as AA, exert a potent social influence and can be used to enhance the effectiveness of other treatment modalities.

Self-help groups are an especially valuable support. Twelve Step groups often provide the client with the most comprehensive network of community support on discharge. Not only can an individual find an AA or NA meeting in most places at most times of day, but the system offers those interested a *sponsor* to help and advise them as they attempt to become abstinent from their addiction. The sponsor is someone who has been in recovery (or not drinking or using drugs) for some time. They can usually be contacted at any time of day or night for advice or to help overcome craving or desire to return to the addictive behaviour. Family members can also find support through groups such as Al-Anon, Families Anonymous or Al-Ateen (groups linked to AA but for family members).

One of the key tasks of rehabilitation will be realignment of support systems away from those that facilitate addictive habits, including lifestyle, involvement in a previous subculture or maladaptive family coping.

Precursors to long-term rehabilitation

In general, the initial priorities for addictions rehabilitation can be placed under three headings:

1   harm minimization;
2   health education and promotion;
3   facilitating change.

Harm reduction

Harm reduction (Single 1995), as the term implies, means helping the addict to learn how to carry out their habit safely, with reduced harm to themselves or others. On first contact with treatment services, many drug users are suspicious and unwilling to give up their addictive behaviour. Experienced drug workers say that attempts to impose this generally drive the client away from the help they need. Harm minimization tends to build confidence in treatment services and can be a step towards changing to reduced drug-using behaviour.

Harm minimization approaches include teaching safe injecting (such as not sharing needles to prevent spread of HIV or hepatitis C), keeping track of alcohol consumption using a 'drink diary' or smoking a lower tar brand of cigarettes. Transfer to a less harmful variant of a drug can minimize harm (such as substitution of buprenorphine for heroin), how to sniff volatile substances safely (e.g. avoidance of aerosols, not sniffing in dangerous environments).

Health education and promotion

As a step further on from harm minimization, health education and promotion can occur when a client is ready and interested in the further effects and health consequences of different drugs. This may occur during a visit to a clinic to dispose of used needles or to collect a methadone script, where the health professional provides a concentrated piece of information relevant to that particular client. Education may occur more formally within a treatment setting, following detoxification, such as a package of information about the effects of alcohol on the individual – taught over several group sessions by an occupational therapist.

The importance of education lies in helping the client to become aware that harmful consequences of their drug use exist and *that they can do something* about them. Encouraging knowledge about drugs can help the client think. Thought about harmful drug effects is usually the precursor to a desire to change drug-using behaviour.

Facilitation of change

If a client indicates readiness for change, here rehabilitation can begin in earnest. It may have taken several admissions to hospital or relapse to harmful drug use to move their thinking to a state where change is desirable.

In order to maximize the effect, an important precursor is the establishment of stable internal motivation for recovery. Initiation of this stable state first requires agreement to enter formal treatment and accept all it involves. Failure to do this will usually mean instant discharge, usually with the offer of treatment at a later time. Once treatment is accepted, a variety of support mechanisms and environmental changes can begin.

The value of groupwork, both task orientated and person centred, is especially valuable in facilitating, supporting and maintaining change (McVinney 1997). The creative opportunities provided by such groups are as wide-ranging as the imagination of therapists; for example, we can use role play within 'rehabilitation courses' for drink-drive offenders (Alcohol Advice Centre 1998), or design 'marathon groups' to promote spiritual development (Page and Berkow 1998), or boost the energy of drug users in a cramped women's prison with dance and karaoke groupwork (Goodison and Schafer 1999). For clients with a long history of drug-related disability, therapeutic communities (Tweedell 1999) can provide a fruitful environment for groupwork aimed at 'changing careers' in the lifestyle of the resident.

Maintenance of lifestyle balance

One of the key points from the 'Programme of Action' in the government's recent strategy recommended that drug users were supported in '... reviewing and changing their behaviour towards more positive lifestyles – linking up where appropriate with accommodation, education and employment services' (Central Drugs Coordination Unit 1998). The European Social Fund is currently supporting work in four countries to link alcohol and drug users to education and employment opportunities (Teller 2000).

One of the most important aspects of rehabilitation is the need to facilitate balance in aspects of a client's lifestyle (Rotert 1990). Part of this often involves assisting the person concerned to restructure their habits and routines. Four aspects of life change stand out:

1 the need to learn or relearn skills for coping and daily living;
2 the need to build and develop new support systems;
3 the need to establish a new life structure, new habits and routines;
4 the need to maintain a balance in aspects of daily living.

These tasks need to be carried out in relation to all aspects of living for rehabilitation to be a long-term success. A useful way of examining the aspects of *lifestyle (or 'way of life')* is to divide it into *leisure and recreation, work and productivity* and *self-maintenance.*

Leisure and recreation

Consideration of leisure needs are a significant aspect of rehabilitation. *It is recommended that leisure is considered first*, before attempting to rehabilitate an individual into any form of work. Leisure is where the individual who is addicted is most likely to relapse. The reasons for this are extensive but mainly relate to the social and physical contexts where the client is most likely to use a particular drug.

Several writers and researchers have identified leisure as being especially maladaptive in alcoholics. Not only do they say alcoholics lose balance in their leisure behaviour, with activities becoming increasingly centred around drinking (as described in the dependence syndrome: Royal College of Psychiatrists 1987), but alcoholics appear to lose the ability to experience positively activities which could be termed leisure related or recreational. Alcoholics Anonymous strongly suggests involvement in constructive leisure activities to counteract 'drinking thinking' associated with an unoccupied mind. As early as 1952, occupational therapy literature about alcoholism and other drug use suggested engagement in leisure as an important aspect of treatment to help the client gain or regain the ability to structure their day (Hossack 1952) and this view is reinforced in recent literature (Rotert 1990; Mann and Talty 1991; Chacksfield and Forshaw 1997). The significant attraction of the large amounts of money that can be earned by drug dealing should not be underestimated, as this can lead to relapse and failed rehabilitation attempts.

Work and 'productivity'

Substance users often have great difficulty maintaining a job or returning to work following treatment. For some, according to the Royal College of Psychiatrists (1987: 36):

> The slog and 'hassle' of ensuring one's drug supply can itself become meaningful; it is something to do, the framework for the day, a job in its own right.

Changing the focus of this kind of routine to a non-drug-related form of work, with the same sense of meaning, can be one of the biggest challenges faced by a drug user in recovery. British culture values the work ethic, and for this reason many addicts view work as a source of meaning and wish to rush back to a job immediately following recovery, apart from the obvious financial incentive and social merit derived from having a work role. In the author's experience, it is important to grade return to work, commencing with perhaps voluntary work until a stable routine has been established, progressing to a part-time job and eventually full time. This can be difficult for someone who needs to earn money. Some employers will facilitate a gradual reintroduction to work as part of rehabilitation.

The concept of work can be extended to mean any form of productive activity which brings a definite reward and which is not done primarily for pleasure (Baum and Christiansen 1997).

Self-care/Self-maintenance

Lack or atrophy of self-care skills can cause significant difficulties for the recovering user. A study by Caan and Fenton (1990) found that adult drug users in a clinical inpatient population were highly likely never to have learned to budget and cook for themselves. Learning these skills can not only facilitate a development of a sense of self-esteem and control but can also encourage confidence and independent living as well as an awareness of healthy eating principles.

Conclusion

Rehabilitation of the biopsychosocial set of problems presented by someone dependent on a substance is clearly a complex business and will invariably take a long period of time. A range of tools and techniques have been evolved to address addictions rehabilitation through the work of the many and varied professionals working in the field. It must be borne in mind that every case is different and success comes with an eclectic approach and an open mind. No one professional group or treatment technique will provide the ultimate answer. Progression is a slow, long haul and sometimes frustrating for both client and professional. However, it is often the challenge of rehabilitation which makes the job highly rewarding to both of these people – especially when success finally comes. It is the subtle and varied nature of the work which makes many people 'addicted' to working in this specialist clinical field.

Discussion points

- Do alcoholics reinforce the disease model by repeatedly telling themselves 'I am an alcoholic'? Is this harmful?
- What are the priorities in addictions rehabilitation – change, situational coping or self-efficacy?
- Is rehabilitation more effective in hospital or in the community?
- Should addicts be rehabilitated in treatment centres over a long period of time or as outpatients in the community?
- Is the approach used by AA and NA outdated?
- Is the impact of addictive behaviour on a person's daily life an important aspect for consideration in rehabilitation? If so, how?
- Should specific professionals carry out specific roles in treatment?

Recommended reading

Alcoholics Anonymous (1975) *Living Sober: Some Methods AA Members Have Used for Not Drinking*. London: Alcoholics Anonymous.

Edwards, G., Marshall, J. and Cook, C. (eds) (1997) *The Treatment of Drinking Problems*, 3rd edn. Oxford: Blackwell Scientific.

Grant, M. and Hodgson, R. (eds) (1991) *Responding to Drug and Alcohol Problems in the Community – A Manual for Primary Healthcare Workers with Guidelines for Trainers*. Geneva: World Health Organization.

Jarvis, T. J., Tebutt, J. and Mattick, R. D. (1995) *Treatment Approaches for Alcohol and Drug Dependence: An Introductory Guide*, Chichester: Wiley.

Royal College of Psychiatrists (1987) *Drug Scenes: A Report on Drugs and Drug Dependence*. London: Royal College of Psychiatrists.

Acknowledgements

My thanks go to Mick MacDaid, Griffith Edwards and David Forshaw for everything they taught me about clinical practice in this area.

References

Alcohol Advice Centre (1998) *Rehabilitation Courses for Drink-Drive Offenders*. Watford: AAC.

Appleby, L. (2001) *Safer Services*, the 5-year report of the confidential enquiry into homicides and suicides. London: Department of Health.

Baum, C. and Christiansen, C. (1997) 'The occupational therapy context: Philosophy, principles – practice', in: C. Christiansen and C. Baum (eds) *Occupational Therapy: Enabling Function and Well-Being*, pp. 27–45. Thorofare, NJ: Slack.

Bennett, W. J., Diiulio, J. J. and Walters, J. P. (1996) *Body Count: Moral Poverty and How to Win America's War Against Crime and Drugs*. New York: Simon and Shuster.

Berg and Miller (1996) *Solution Focused Therapy*.

Birmingham, L. (1999) 'Between prison and the community', *Brit. J. Psychiat.*, **174**, 378–9.

Brown, C. and Brooke, D. (1991) 'Protecting your assets: Caring for the therapist', in I. Glass (ed.) *The International Handbook of Addiction Behaviour*, pp. 269–73. London: Routledge.

Caan, W. and Crowe, M. (1994) 'Using readmission rates as indicators of outcome in comparing psychiatric services', *J. Mental Health*, **3**, 521–4.

Caan, W. and Fenton, J. (1990) 'Self-catering during rehabilitation', *Brit. J. Psychiat.*, **157**, 780–1.

CAOT (1993) *Guidelines for Client-Centred Mental Health Practice*. Toronto: Canadian Association of Occupational Therapists in Mental Health.

Central Drugs Coordination Unit (1998) *Tackling Drugs to Build a Better Britain*. London: Her Majesty's Stationery Office.

Chacksfield, J. D. and Forshaw, D. M. (1997) 'Occupational therapy and forensic addictive behaviours', *Brit. J. Therapy Rehabilitation*, **4**, 381–95.

Cutland, L. (1990) *Freedom from the Bottle: A Guide to Recovery for Alcoholics, Their Partners and Children*. Bath: Gateway Books.

Davies, J. B. (1997) *Drugspeak: The Analysis of Drug Discourse*. Amsterdam: Harwood Academic.

Davison, G. C. and Neale, J. M. (1990) *Abnormal Psychology, 5th edn*. New York: Wiley.

Dittrich, J. E. (1993) 'A group program for wives of treatment-resistant alcoholics', in T. J. O'Farrell (ed.) *Treating Alcohol Problems: Marital and Family Interventions*. New York: Guilford Press.

D'Orban, P. T. (1986) 'Drugs and alcohol: The psychiatrist as expert witness in court', *Brit. J. Addict.*, **81**, 631–9.

Driessen, M., Veltrup, C., Weber, J., John, O., Wetterling, T. and Dillon, H. (1998) 'Psychiatric co-morbidity, suicidal behaviour and suicidal ideation in alcoholics seeking treatment', *Addiction*, **93**, 889–94.

Drummond, D. C., Tiffany, S. T., Glautier, S. and Remington, B. (eds) (1995) *Addictive Behaviour: Cue Exposure Theory and Practice*. Chichester: Wiley.

Edwards, G. (ed.) (1987) *The Treatment of Drinking Problems*, 2nd edn. Oxford: Blackwell Scientific.

Edwards, G. and Gross, P. (1976) 'Alcohol dependence: Provisional description of a clinical syndrome', *Brit. Med. J.*, **1**(6017), 1058–61.

Edwards, G., Brown, D., Duckitt, A., Oppenheimer, E., Sheehan, M. and Taylor, C. (1987) 'Outcome of alcoholism: The structure of patient attributions as to what causes change', *Brit. J. Addict.*, **82**, 533–45.

Edwards, G., Oppenheimer, E. and Taylor, C. (1992) 'Hearing the noise in the system: Exploration of textual analysis as a method for studying change in drinking behaviour', *Brit. J. Addict.*, **87**, 73–81.

Ellis, A., McInerny, J. F., DiGiuseppe, R. and Yeager, R. J. (1988) *Rational Emotive Therapy with Alcoholics and Substance Abusers*. New York: Pergamon Press.

Fordham, F. (1966) *An Introduction to Jung's Psychology, 3rd edn*. Harmondsworth: Penguin.

Foster, J. H., Powell, J. E., Marshall, E. J. and Peters, T. J. (1999) 'Quality of life in alcohol-dependent subjects – a review. *Quality of Life Research*, **8**, 255–61.

Gafoor, M. and Rassool, G. H. (1998) 'Working with dual diagnosis clients', in G. H. Rassool (ed.) *Substance Use and Misuse: Nature, Context and Clinical Interventions*, pp. 249–59. Oxford: Blackwell Science.

Galanter, M. (1987) 'The large-group social support network in the treatment of substance abuse: Naturalistic and experimental models', in: E. Gottheil, K. A. Druley, S. Pashko and S. P. Weinstein (eds) *Stress and Addiction*. New York: Brunner/Mazel.

Ghodse, H. (1989) *Drugs and Addictive Behaviour: A Guide to Treatment*. Oxford: Blackwell Scientific.

Goldstein (1986) 'The drugs/violence nexus: A tripartite conceptual framework', *J. Drug Issues*, **15**, 493–506.

Goodison, L. and Schafer, H. (1999) 'A dance to the music of time', *Health Service J.*, 21 October 1999, 28–9.

Glass, I. (ed.) (1991) 'Career and natural history', in *The International Handbook of Addiction Behaviour*, pp. 34–40. London: Routledge.

Gossop, M. (ed.) (1989) *Relapse and Addictive Behaviour*. London: Routledge.

Gossop, M. (1994) 'Drug and alcohol problems: Treatment', in S. J. E. Lindsay and G. E. Powell (eds) *Clinical Adult Psychology, 2nd edn*, pp. 384–412. London: Routledge.

Henderson, S. (1997) *Ecstasy: Case Unsolved*. London: Pandora.

Hetherington, C. (1995) 'Dysfunctional relationship patterns: Positive changes for gay and lesbian people', in R. J. Kus (ed.) *Addiction and Recovery in Gay and Lesbian Persons*. New York: Haworth Press.

Holder, J. and Williams, T. (1995) *Perceptual Adjustment Therapy: A Positive Approach to Addictions Treatment*. Washington, DC: Accelerated Development.

Holland, M. (1998) 'Substance use and mental health problems: Meeting the challenge', *Brit. J. Nurs.*, **7**, 896–900.

Hossack, J. R. (1952) 'Clinical trial of occupational therapy in the treatment of alcohol addiction', *Am. J. Occupational Therapy*, **6**, 265–82.

Kaufman, E. (1985) *Substance Abuse and Family Therapy*. Orlando, FL: Grune & Stratton.

Kearney, R. J. (1996) *Within the Wall of Denial: Conquering Addictive Behaviours*, London: W. W. Norton.

Kent, R. (1990) 'Focusing on women', in S. Collins (ed.) *Alcohol, Social Work and Helping*. London: Routledge.

Kielhofner, G. (1985) *The Model of Human Occupation*. Baltimore, MD: Williams & Wilkins.

Kielhofner, G. (1995) *The Model of Human Occupation*, 2nd edition. Baltimore, MD: Williams and Wilkins.

Kivlahan, D. R., Donovan, D. M. and Walker, R. D. (1983) 'Predictors of relapse: Interaction of drinking-related locus of control and reasons for drinking', *Addict. Behav.*, **8**(3), 273–6.

Leckie, T. (1990) 'Social work and alcohol', in S. Collins (ed.) *Alcohol, Social Work and Helping*. London: Routledge.

Lesieur, H. L. and Blume, S. B. (1988) 'The South Oaks Gambling Screen (SOGS): A new instrument for the identification of pathological gamblers', *Amer. J. Psychiat.*, **144**, 1184–8.

Levine, B. and Gallogy, V. (1985) *Group Therapy with Alcoholics: Outpatient and Inpatient Approaches*. London: Sage.

McColl, M. A., Gerein, N. and Valentine, F. (1997) 'Meeting the challenges of disability: Models for enabling function and well being', in C. Christiansen and C. Baum (eds) *Occupational Therapy: Enabling Function and Well-being*, pp. 509–28. Thorofare, NJ: Slack.

McLellan, A. T., Hagan, T. A., Levine, M., Gould, F., Meyers, K., Bencivengo, M. and Durell, J. (1998) 'Supplemental social services improve outcomes in public addiction treatment', *Addiction*, **93**, 1489–99.

McVinney, D. L. (1997) *Chemical Dependency Treatment: Innovative Group Approaches*. Binghampton, NY: Haworth Press.

Mann, W. C. and Talty, P. (1991) 'Leisure activity profile: Measuring use of leisure time by persons with alcoholism', *Occupational Therapy in Mental Health*, **10**(4), 31–41.

Marlatt, G. A. and Gordon, J. R. (1985) *Relapse Prevention: Maintenance Strategies in the Treatment of Addictive Behaviours*. New York: Guilford.

Miller, W. R. (1983) 'Motivational interviewing with problem drinkers', *Behav. Psychother.*, **11**, 147–72.

Miller, W. R. and Rollnick, S. R. (1991) *Motivational Interviewing: Preparing People to Change Addictive Behaviour*. New York: Plenum Press.

Minshull, J., Ross, K. and Turner, J. (1986) 'The human needs model of nursing', *J. Adv. Nurs.*, **11**(6), 643–9.

Moos, R. H. (1993) *Coping Responses Inventory. Professional manual*. Odessa, FL: Psychological Assessment Resources.

Moos, R. H., Bromet, E., Tsu, V. and Moos, B. (1979) 'Family characteristics and the outcome of treatment for alcoholism', *J. Stud. Alcohol*, **40**(1), 78–88.

O'Farrell, T. J. (ed.) (1993) *Treating Alcohol Problems: Marital and Family Interventions*. New York: Guilford Press.

Page, R. C. and Berkow, D. N. (1998) 'Group work as facilitation of spiritual development for drug and alcohol abusers', *J. Specialists in Group Work*, **23**, 285–97.

Prochaska, J. O. and DiClemente, C. C. (1983) 'Stages and processes of self-change of smoking: Toward an integrative model of change', *J. Consult. Clin. Psychol.*, **51**, 390–5.

Prochaska, J. O. and DiClemente, C. C. (1986) 'Toward a comprehensive model of change', in W. Miller and N. Heather (eds) (1986) *Treating Addictive Behaviours*. New York: Plenum Press.

Raistrick, D. (1991) 'Career and natural history', in I. Glass (ed.) *The International Handbook of Addiction Behaviour*, pp. 34–40. London: Routledge.

Rosenthal, M. S. (1991) 'Therapeutic communities', in I. Glass (ed.) *The International Handbook of Addiction Behaviour*, pp. 258–63. London: Routledge.

Rosenthal, R. N. and Westreich, L. (1999) 'Treatment of persons with dual diagnoses of substance use disorder and other psychological problems', in B. S. McCrady and E. E. Epstein (eds) *Addictions: A Comprehensive Guidebook*, pp. 439–76. New York: Oxford University Press.

Rotert, D. A. (1990) 'Occupational therapy and alcoholism', *Occupational Med.*, **4**, 327–37.

Rotter, J. B. (1966) 'Generalized expectancies for internal versus external locus of control of reinforcement', *Psychological Monographs*, **80**, 1–28.

Royal College of Psychiatrists (1986) *Alcohol: Our Favourite Drug*. London: Tavistock.

Royal College of Psychiatrists (1987) *Drug Scenes: A Report on Drugs and Drug Dependence*. London: Royal College of Psychiatrists.

Ryle, A. (1991) *Cognitive Analytic Therapy: Active Participation in Change*. Chichester: Wiley.

Siegal, S. (1999) 'Drug anticipation and drug addiction: The 1998 H. David Archibald lecture', *Addiction*, **94**, 1113–24.

Single, E. (1995) 'Defining harm reduction', *Drug and Alcohol Rev.*, **14**(3), 287–90.

Skinner, B. F. (1978) *About Behaviourism*. London: Penguin.

Smart, R. (1994) 'Dependence and the correlates of change: A review of the literature', in G. Edwards and M. Lader (eds) *Addiction: Processes of Change*, pp. 79–94. Oxford: Oxford University Press.

Sobell, L. C., Sobell, M. B., Toneatto, T. and Leo, G. I. (1993) 'What triggers the resolution of alcohol problems without treatment?', *Alcoholism; Clin. Experiment. Res.*, **17**(2), 217–24.

Steinglass, P., Bennet, A., Wolin, S. J. and Reiss, D. (1987) *The Alcoholic Family: Drinking Problems in a Family Context*. London: Hutchinson.

Swanson, J. W., Holzer, C. E., Ganju, V. K. and Jono, R. T. (1990) 'Violence and psychiatric disorder in the community: Evidence from the epidemiologic catchment area surveys', *Hosp. Community Psychiatry*, **41**, 761–70.

Taylor, B. and Bennett, T. (1999) *Comparing Drug Use Rates of Detained Arrestees in the United States and England*. Washington, DC: National Institute of Justice.

Teller, J. (2000) 'Drug and alcohol initiative in Avon', *NHS Magazine*, **20**, 21.

Tweedell, D. (1999) 'Therapeutic communities helped people to recover from substance abuse and implement "new lives"', *Evidence-Based Nursing*, **2**, 28.

Velleman, R. (1992) *Counselling for Alcohol Problems*. London: Sage.

Willenbring, M. L. (2000) 'Integrated outpatient treatment increased abstinence in men with alcohol related illness who were ongoing drinkers', *Evidence-Based Mental Health*, **3**, 54.

Chapter 18

Policing drug abuse

Graham Divall

Key points

- *In each country, drug use occurs within a specific legal context.*
- *In the UK, legislation is related to the type of drug and its intended use (e.g. intent to produce or supply an illicit substance).*
- *It therefore becomes important to identify different substances accurately, including the source and typical usage of each drug encountered.*

Drugs and the law

The Misuse of Drugs Act

The most important law for controlling drugs in the UK is *The Misuse of Drugs Act 1971* together with *Misuse of Drugs Regulations* made under that Act.

Police powers

The Misuse of Drugs Act 1971 (MUDA) provides the police with powers to:

- stop and search a person or vehicle;
- obtain search warrants to search premises;
- seize anything which appears to be evidence of an offence;
- arrest a person suspected of having committed an offence.

Offences

Under the provisions of the MUDA, it is illegal:

- to possess a controlled drug;
- to possess with intent to supply a controlled drug;
- to supply a controlled drug;
- to produce a controlled drug;

- for an occupier or person in charge of any premises to allow anyone to produce, supply or offer to supply controlled drugs;
- to supply or offer to supply articles used in the administration or preparation for administration of controlled drugs.

Classes of drug

The MUDA divides controlled drugs into three classes (A, B and C):

- *Class A drugs* are considered to be the most harmful or dangerous, and include heroin (diamorphine), methadone, cocaine, LSD (lysergic acid diethylamide), and MDMA (methylenedioxymethamphetamine). The maximum penalty for supplying a class A drug is life imprisonment and 7 years imprisonment for possession.
- *Class B drugs* include amphetamine, methylamphetamine, cannabis, cannabis resin, dihydrocodeine and barbiturates. The maximum penalties are 14 years imprisonment for supply and 5 years for possession.
- *Class C drugs* include most benzodiazepines, phentermine and buprenorphine. The maximum penalty for supply is 5 years imprisonment and 2 years for possession.

Dealing with new substances

When considering the legal control of drugs, many countries have relied upon specific chemical listings. An important issue here is that when a new substance of abuse is created or manufactured that substance has to be added to the list; for example, when the USA made MDA illegal, people started to use MDMA until it was listed. In Holland, when MDMA became illegal, MDEA continued to be legal until the law was changed again.

The UK has developed a more flexible approach by adopting generic definitions of scheduled substances and thus avoids having to change the lists of controlled substances every time a new chemical variant appears. This approach does not cover every eventuality and the Misuse of Drugs Act is kept under review by the Advisory Council on the Misuse of Drugs (ACMD).

Other important legislation

The Medicines Act 1968

This controls the manufacture and supply of medicines. It divides medicinal products into three categories:

- *Prescription-only medicines* can only be obtained on a doctor's prescription and can only be supplied at a registered pharmacy under the supervision of a registered pharmacist (e.g. benzodiazepines).
- *Pharmacy medicines* can only be supplied and sold at a registered pharmacy under the supervision of a registered pharmacist (e.g. painkilling preparations containing dihydrocodeine and codeine).
- *The General Sales List* contains the least harmful medicinal products and these can be supplied and sold in small quantities by shops and supermarkets.

The Intoxicating Substances [Supply] Act 1985

Many substances not controlled by the MUDA have been abused. These include glues, thinners, solvents and butane gas. This Act makes it illegal to supply to people under the age of 18 years substances which cause intoxication when inhaled and when the supplier knows, or has cause to believe, that the substance is likely to be used for intoxication.

The Criminal Justice International Co-operation Act 1990

This controls the importation and exportation of chemicals which could be used for the illegal manufacture of drugs.

The Road Traffic Act 1988

It is an offence to drive or attempt to drive a motor vehicle on a road or public place when one's ability to drive properly is impaired by alcohol or drugs. In this context, drugs includes medicines, such as insulin for diabetics. The Act specifies prescribed limits for alcohol and these are:

- 35 μg of alcohol in 100 ml of breath;
- 80 mg of alcohol in 100 ml of blood;
- 107 mg of alcohol in 100 ml of urine.

Breath alcohol levels for evidential purposes can be measured by trained police officers using Home Office approved instruments. Levels of alcohol in blood and urine require laboratory analysis by authorized analysts.

There are no prescribed limits for drugs. A charge is usually proved by evidence of:

- impairment (e.g. findings of a doctor during medical examination and observations of a police officer);
- the presence of a drug or breakdown product (metabolite) in the driver's blood or urine as determined by laboratory analysis.

The Road Traffic Act 1991

This Act makes it a specific offence to cause death by careless driving when under the influence of drink or drugs.

The Transport and Works Act 1992

This Act makes it an offence for a person to work on a transport system (e.g. drivers, guards, conductors, signalmen, maintenance crew and supervisors), when that person is unfit to carry out that work through drink or drugs.

The Crime and Disorder Act 1998

This deals with a wide range of matters concerning the prosecution and processing of criminal offences. One part of the Act makes provision for rehabilitation of repeated offenders who commit crimes to support and finance a drug-taking habit. It is proposed that persons will be required to attend a drug education and drug testing programme. The implications of these new Drug Treatment and Testing Orders for health and social care practitioners have been described recently by Quinn and Barton (2000).

Main substances of abuse

Heroin

Source

Heroin (diamorphine) is derived from the opium poppy (*Papaver somniferum*). Areas where it is grown and processed include Pakistan, India, Afghanistan, the Far East, Iran and Turkey. Approximately 85–90 per cent of the heroin reaching Europe is from Western Asia. Pharmaceutical heroin is also manufactured. It is one of the most powerful analgesics available.

Appearance

In the UK, illicit heroin is seen as a light- to mid-brown powder (Plate 18A). In larger quantities, it is found as compressed powder blocks contained in distinctive wrappings. Currently, the majority of illicit heroin in the UK has a purity of between 10 and 65 per cent. Common 'cutting agents' include caffeine and paracetamol. Pharmaceutical heroin is prepared as white tablets or freeze-dried powder in glass ampoules.

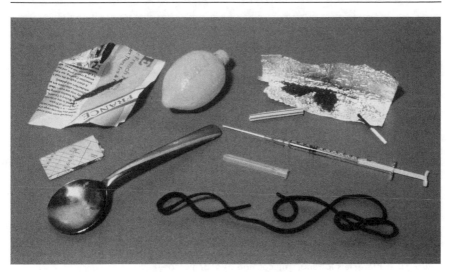

Figure 18.1 Materials used for the abuse of heroin.

Method of use

Heroin is dissolved in water with the aid of weak acid such as lemon juice or vitamin C and then injected beneath the skin ('popping') or directly into veins ('mainlining'). Associated materials can include matches, spoons, hypodermic syringes, needles and tourniquets (Figure 18.1).

The substance is also smoked with tobacco. A common smoking method is to place the heroin on a piece of tin foil and heat it directly and inhale the curling plumes of smoke through a straw ('chasing the dragon').

Cocaine

Source

Cocaine is a stimulant drug derived from the leaf of the coca plant (*Erythroxylum coca*) which grows in the mountainous regions of South America – Colombia, Brazil, Peru and Bolivia.

Appearance

Cocaine exists as either the hydrochloride salt or as a base. The salt appears as a white crystalline powder (Plate 18B). The purity of cocaine hydrochloride at street level varies enormously. Current data indicates that the purity of imported cocaine hydrochloride is somewhat higher and is cut or diluted with substances such as lignocaine, caffeine, glucose and other sugars.

Figure 18.2 Materials for snorting cocaine hydrochloride.

Cocaine hydrochloride is often converted to the base which, depending on the method used, appears as a whitish, waxy material or as whitish–beige rocky lumps known as 'crack' (Plate 18C). Crack is not a new drug, as reported in the Press a few years ago, but is chemically identical to cocaine base. Crack is usually between 80 and 95 per cent pure.

Methods of use

Cocaine base is volatile when heated and is abused by smoking using home-made pipes made from soft-drink cans, bottles and glass tubing. The drug is heated in the pipe using matches or a lighter and the fumes are inhaled ('freebasing'). Purpose-made glass pipes are also (illegally) available.

Cocaine hydrochloride cannot be smoked because it decomposes when heated. Rather, it is abused by snorting or sniffing the drug. It is usually placed on a smooth surface such as a mirror, finely divided and made into a thin line using a blade and then snorted through a straw or rolled banknote or similar tube (Figure 18.2). Cocaine is also injected and can be mixed with heroin ('speedballing').

LSD

Source

LSD or lysergic acid diethylamide is a powerful hallucinogenic drug produced synthetically from precursor chemicals. The main sites of production appear to be the western states of the USA.

Plate 18A Illicit heroin.

Plate 18B Cocaine hydrochloride.

Plate 18C Cocaine base.

Plate 18D LSD paper squares.

Plate 18E Herbal cannabis.

Plate 18F Cannabis resin.

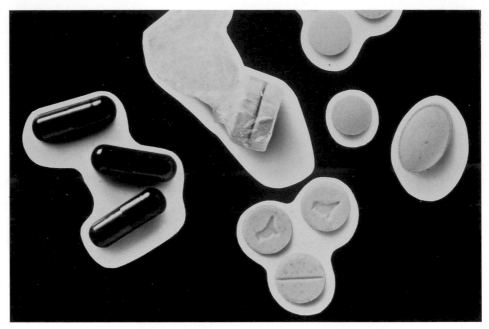

Plate 18G Ecstasy tablets and capsules.

A man aged 40 who misused drugs and had had a leg amputated after ischaemic damage from intra-arterial injections presented with blindness of recent onset. He was blind in both eyes. The left eye was ophthalmoplegic, with corneal clouding and no pupillary reflexes. This was the result of his injecting gel temazepam into the inner canthus. This substance is known to cause vascular occlusion.
—J L THOMPSON, *senior house officer in neurology,* D HONEYBOURNE, *consultant physician,* R E FERNER, *consultant physician, Dudley Road Hospital, Birmingham B18 7QH.*

Plate 18H Damage to eye after injecting gel temazepam (see p.206).

Appearance

LSD first became popular in the hippie culture of the 1960s. It has appeared in many forms including capsules, gelatine squares, impregnated sugar cubes, microdots and paper squares. The latter is the current form of availability. The drug is absorbed by dipping or spotting onto paper squares (approximately $5 \times 5\,mm$) which bare distinct markings or motifs such as cartoon characters and shapes (Plate 18D).

Method of use

LSD is ingested orally. Each square is a 'dose' and contains as little as 25 µg of the drug. Hallucinogenic effects begin after about 30 min. Some users who have not taken the drug recently experience hallucinations know as 'flashbacks'

MDMA

Source

MDMA or 'Ecstasy' is a drug with stimulant, mood altering and sometimes hallucinogenic effects. It is manufactured from precursor chemicals. Most of the drug is imported, although several illicit laboratories have been found in the UK. Other drugs which are chemically similar and have similar effects have also been synthesized. The most common are:

- MDA (methylenedioxyamphetamine);
- MDEA (methylenedioxyethylamphetamine);
- MBDB (methylbenzodioxybutanamine).

Ecstasy is now a generic term and only laboratory analysis can distinguish between these closely related substances.

Appearance

Ecstasy is most frequently encountered as white, off-white, beige, pale pink or pale blue tablets, about the size of pharmaceutical aspirin or paracetamol, which often bear a characteristic motif or logo such as a dove. The drug has also been found in capsules (Plate 18G).

Method of use

Ecstasy is ingested orally and is most closely associated with the emergence of the 'acid house' and 'rave' culture.

Figure 18.3 Amphetamine powder in a paper wrap.

Amphetamine

Source

Amphetamine was first synthesized in Germany in the 1880s. It is a stimulant drug and became a popular drug of abuse in the 1960s when it was available as Drinamyl tablets, also known as 'purple hearts' because of their shape and colour. It has also been prescribed as an appetite suppressant and to treat narcolepsy. Illicit forms of the drug are synthesized from reasonably available precursor chemicals and several methods/synthetic routes are used. The drug is imported into the UK and also manufactured here in illicit laboratories.

Appearance

Amphetamine is found as white, off-white, pale yellow, green or pink tablets and as a powder in a variety of colours and textures (Figure 18.3). The drug contains minor impurities which can be detected by laboratory analysis and is the basis of linking amphetamine seizures. At the street level, it is usually between 2 and 10 per cent pure having been cut with one or more bulking agents, the most common being sugars such as lactose, glucose and mannitol.

Method of use

Amphetamine is ingested orally and can be dissolved in drinks. It is sometimes injected or snorted.

Cannabis

Source

Cannabis is derived from the plant *Cannabis sativa* and is encountered in a variety of forms:

- the growing plant;
- herbal material;
- cannabis resin;
- hash oil.

The plant is grown widely in Africa, Asia, the Middle and Far East, and the Americas. It is also grown in the UK but normally only reaches full maturity if grown indoors. It is increasingly cultivated by hydroponics. This is where the plant is grown without soil using water, nutrients, inert support materials and with carefully controlled temperatures and lighting conditions (Figure 18.4). High-quality cannabis known as 'skunk' can be produced by this method.

All parts of the cannabis plant contain a group of compounds known as cannabinoids. The highest concentration of the active constituent THC (tetrahydocannabinol) is found in the fruiting tops of the mature plant (Figure 18.5). This part of the plant is harvested as cannabis or used to produce cannabis resin. These are the two common forms commonly imported into the UK. The whole of the cannabis plant above the ground, except for seeds and stalks which are separate from the plant, falls within the legal definition of cannabis.

Appearance

The cannabis plant may grow to 10–15 feet in the right environment. Its leaves are formed by an odd number of saw-edged leaflets. Typically, herbal cannabis or 'marijuana' consists of chopped dried plant material (Plate 18E).

Cannabis resin or 'hashish' is a brown, sticky, resinous substance produced by hairs which cover the cannabis plant. This is collected and compressed into blocks which may be soft and pliable or dry and hard (Plate 18F). The physical appearance of cannabis resin differs depending on the part of the world from which it originates. Resin from Asia is dark

Figure 18.4 Cannabis plants cultivated by hydroponics.

brown or black whilst Moroccan resin occurs as either greenish brown slabs or as black slabs the size of a bar of soap.

When cannabis resin is extracted with a solvent such as alcohol and the solvent is allowed to evaporate a dark green/brown tarry substance known as 'hash oil' is formed. This has a high concentration of THC.

Figure 18.5 Fruiting top of the cannabis plant.

Methods of use

Cannabis is commonly consumed by smoking, which entails the herbal material or resin being used in a pipe or incorporated into a hand-rolled cigarette usually mixed with tobacco (a 'reefer' or 'joint').

In some criminal defences, it is sometimes claimed that the presence of cannabinoids in a blood sample has resulted from the secondary inhalation of cannabis smoke by the subject.

Cannabis is also ingested orally and is found incorporated into cakes and other recipes.

Drug trafficking and multiple sources

Since it depends on the changeable political and economic links between countries, the pattern of international trade in drugs can change over time. Policing this illicit trade needs continuous vigilance and international co-operation to respond to the multiple sources and transport routes used for a drug like heroin (Keep 1998).

Scientific services and the MUDA

A range of scientific services and tests are required to support investigations and charges brought under the MUDA. These vary in complexity depending on the nature of the charge and include the following.

Corroborative testing

When a person admits to being in possession of a small quantity of cannabis, amphetamine, morphine or diamorphine, then some form of corroborative testing is allowed rather than unequivocal identification by laboratory analysis.

In the case of cannabis, a visual identification by a police officer is all that is required. For amphetamine, morphine and diamorphine, a trained police officer can apply a simple spot chemical test (the Marquis test) which generates a recognizable colour with each of the drugs. Test kits for this purpose have to be authorized by the Home Office.

Currently, cocaine and the Ecstasy drugs cannot be tested in this way. However, consideration is now being given to extending the types of test kit which can be used and the range of drugs which can be tested.

Identification and purity determination

Unequivocal identification of a controlled substance is generally required for all charges brought under the Misuse of Drugs Act. The method of choice for most substances is gas chromatography–mass spectrometry (GC–MS).

Determination of purity is often required where charges relating to supply are involved. For cocaine and morphine, some form of purity determination or estimation is required. This is to demonstrate that the material is controlled by the MUDA because its purity is greater than the level at which possession of material is exempt; for example, preparations containing less than 0.1 per cent cocaine are exempt.

Linking items

When evidence is required to support charges relating to the supply of controlled drugs, then links between items, individuals and places are particularly useful.

Physical comparison

Drugs are often supplied in small wraps of paper, cling film or aluminium foil. When such material is torn or cut, the edges of the separate pieces so obtained will have a close relationship with each other and can be fitted back together again. Such 'physical fits' are generally unique and provide absolute proof that the pieces were once part of the same piece of material.

Sometimes, the physical fit may not be entirely unique but examination can still demonstrate linking features (e.g. between pieces of cannabis resin that have been cut or broken or between pieces of cling film that have been torn and stretched).

In some instances, no physical fit can be found between two pieces of material but other physical characteristics can provide evidence of a relationship between the pieces; for example, paper squares cut from a page of a magazine can be pieced together and shown to be from the same page and one which is missing from a magazine found at the home of a suspected supplier.

Manufacturing defects and features can also be used to link two or more pieces of material; for example, in the manufacture of thin plastic film (used for the manufacture of plastic bags and food wrap material), random stress imperfections and striation marks are induced into the material. These can be detected and compared using polarized light. Similarities or matches between two pieces of such material provide evidence of a relationship and common origin.

Chemical comparison

Illicit drugs are seldom pure and contain minor impurities from their preparation or manufacturing process. Other substances such as bulking agents will be found in different proportions and type. The nature and proportions of these materials provide a means of comparing two drug samples and drawing conclusions about a possible common origin; for example, illicit amphetamine contains impurities which have arisen during its production. The separation and detection of these minor components produces an 'impurity profile'. The profiles can vary widely due to differences in the production method and the laboratory in which the drug was produced. Profiles can be compared to provide a means of linking samples to a common laboratory or to the same manufacturing batch. Such links can still be established even if the original amphetamine has subsequently been cut to a street level purity.

Other evidence of links

Contact trace evidence such as fingerprints, fibres, hairs and body fluids can be used to link items, places and people. A most significant way of doing this in recent years is through DNA profiling of body fluids; for example, deals or wraps of drugs are sometimes secreted in a supplier's mouth. DNA can be recovered from the cellular components of saliva, profiled and compared with samples taken from a suspect.

Drugs and other criminal offences

The previous sections have presented an overview of the law relating to drugs primarily through the provisions of the Misuse of Drugs Act. The law relating to drugs and driving has also been mentioned.

Drugs are also encountered in a wide range of other criminal offences. Drugs and alcohol are used and abused because they are psychoactive. Such substances are often detected in the body samples of both victims and assailants of violent crime. By definition, psychoactive drugs can alter a person's state of mind, affect their mood, memory and state of consciousness. The Criminal Courts often wish to know what drugs had been taken and any such evidence is given due consideration in the Court's deliberations.

In many instances, the taking of drugs is of a secondary nature in relation to the criminal offence but this is not always the case. Some substances are actively administered to a victim to induce sleep or a state of unconsciousness; for example, there has been an increasing incidence of administering flunitrazepam (Rohypnol), a potent sleep-inducing benzodiazepine, to victims of sexual assault. In this way, it has acquired the slang term 'date rape drug'.

It is preferable to have both blood and urine specimens for the detection of drugs in these circumstances. Clearly, other body tissues and fluids can be obtained from deceased persons. It can be hours or days after an incident and, sometimes, it is necessary to identify a drug metabolite rather than the parent compound (e.g. benzoylecgonine is a major metabolite of cocaine). In this way, it is still possible to provide evidence of a drug having been consumed. Samples are usually screened for a range of drugs using enzyme or immuno-based assays. Preliminary results are then confirmed using chromatography and mass spectrometry.

Workplace drug testing

Widespread use and abuse of drugs has become a significant concern to many employers. Absenteeism, poor efficiency and an increased risk of harm and injury in the workplace are major considerations. Many organizations are now developing drug and alcohol policies. In some occupations, drug taking would warrant instant dismissal. In other instances, organizational policy is to support and educate the employee away from drugs. More 'healthy workplaces' became one of the national priorities in the 1999 public health strategy 'Saving Lives: Our Healthier Nation'. Such programmes often require the employee to agree to participate in a drug-testing programme.

A test for your understanding thus far

Based on the provisions of The Misuse of Drugs Act regarding a person in charge of any premises who allows the supply of controlled drugs, two managers of a service for homeless people in Cambridge (Ruth Wyner and John Brock) recently received substantial prison sentences after incidents of clients dealing drugs in the car park of their service. Imagine you are the manager of a small community service (say a Youth Offending Team) likely to come into frequent contact with illicit 'user/dealers' who might sell on drugs in 1 g 'wraps' in cling film:

- How would you write a policy for your staff to prevent the supply of drugs between clients brought together at your service?
- How would you introduce this policy to your clientele?

References

Keep, J. (1998) 'Her Majesty's Customs and Excise: Drug Trafficking', in G. V. Stimson, C. Fitch and A. Judd (eds) *Drug Use in London*, pp. 96–106. London: Leighton Print.

Quinn, C. and Barton, A. (2000) 'The implications of drug treatment and testing orders', *Nursing Standard*, **14**(27), 38–41.

Chapter 19

Prevention

Sue Drummond

Key points

- *Prevention is an essential component of alcohol and drug strategies.*
- *Effective interventions to prevent use and/or harm usually incorporate a package of multiple activities.*
- *Important dimensions in addressing alcohol and drug problems are: control at the population level; reducing harm at the individual level by improving personal skills and resilience; targeting 'high-risk' groups like children in care homes, homeless people or young offenders; community action with high levels of local participation.*
- *Prevention strategies are more likely to succeed if they focus on achievable goals which make sense to people within their everyday life and which gain the support of the wider community.*
- *Prevention initiatives are more likely to succeed if policy makers, professionals and community organizations support them equally.*
- *Under some circumstances, a brief, focused consultation with a familiar professional such as the family doctor can initiate change.*

Throughout your professional careers, you are likely to encounter situations where there is a need to reduce the incidence of alcohol- and drug-related problems; the first section of this chapter examines such a situation in an Accident & Emergency department. Similarly, you are likely to have patients ask you for advice on preventative issues; the second section of this chapter investigates two typical patient enquiries.

Prevention strategies: a case study in an Accident & Emergency department

Imagine you are working in a busy Accident & Emergency (A&E) department, already stretched for resources, when a review of admissions over the past year shows a dramatic increase in the number of alcohol- and

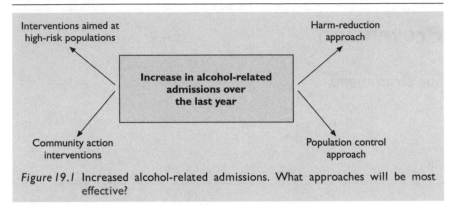

Figure 19.1 Increased alcohol-related admissions. What approaches will be most effective?

drug-related admissions. This is unlikely to be a unique situation as surveys conducted in the UK show that over 10 per cent of people attending A&E departments have been drinking alcohol (Walsh and Macleod 1983) with this figure increasing to up to 40 per cent in the evenings (Holt *et al.* 1980). Approximately 11 per cent of A&E attendees are likely to have *problems* with alcohol and only a very small percentage of these people are likely to have received help for this difficulty. In order to consider what preventative strategy might be best suited to the A&E situation described, it is necessary first to consider the different types of prevention and then the most common prevention approaches. Figure 19.1 shows the approaches most commonly used in the prevention of alcohol and drug problems. Prevention activity is frequently classified as *primary*, *secondary* or *tertiary* prevention. These terms are defined below:

- primary prevention is concerned with preventing the onset of alcohol or drug problems amongst those not using substances, typically children and young people;
- secondary prevention is concerned with preventing people at potential risk from developing problematic drug and or alcohol use;
- tertiary prevention is concerned with preventing people already identi-fied as having harmful alcohol and drug use from developing further problems.

While it is useful to have an understanding of these terms, it is important to remember that prevention activity often combines interventions at two or three levels, together. This will become evident as we next examine the different prevention approaches used in practice. The question is:

Which of these approaches is likely to be most effective in responding to alcohol use in the A&E situation?

Table 19.1 Potential approaches to prevention

	Rationale	Goal	Strategies
Population control strategy	The greater a country's alcohol consumption, the greater the level of alcohol problems	Reduction of the total volume of alcohol consumed by members of one population	Controlling the number and location of alcohol sale premises Licensing regulation Drinking age restriction Laws about serving to intoxicated persons Alcohol taxation Advertising and sponsorship restrictions
Harm reduction approach	Not all drug users will cease or decrease their use at the current time and, so, it is more realistic to reduce harm related to drinking	Reduction of the harm related to drinking	Reduce non-beverage alcohol consumption by homeless Reduce alcohol intake of heavy drinkers Minimize potential harm in drinking places (e.g. shatterproof glasses, padded furniture)
Targeting high-risk populations	The most efficient way to prevent the onset and development of alcohol and drug problems is to target population groups most at risk of developing these problems	Preventing high-risk populations from developing or further developing alcohol and drug problems	Education in schools Intervention in prisons Intervention in A&E departments
Community action approach	Behavioural change is more likely if interventions are focused on changing community norms and practices rather than an individual's behaviour	Change community's norms and practices	Drink-driving campaign Server intervention Media campaigns

To assist you in addressing this question, a summary of the different approaches is given in Table 19.1.

Control strategies aimed at populations

Population control strategies have until recently been the main focus of alcohol prevention policy and practice. However, public support for these strategies has declined with people questioning why heavy alcohol taxes should be imposed because of 'other people's' problems with alcohol. Moreover, restrictions on the sale and price of alcohol are becoming more difficult to enforce, particularly in Europe, with the relaxation of border controls. Recent publicity, albeit often exaggerated, about the potential health benefits of alcohol have made it more difficult to encourage people to drink less. Confusion amongst health professionals and policy makers on what constitutes 'safe' drinking has seen people become sceptical about government advice on safe drinking guidelines. However, simple matters of public policy can affect alcohol-related deaths significantly; for example, raising the minimum legal drinking age is associated with decreased youth suicide (Birckmayer and Hemenway 1999).

Harm reduction approaches

People may be resistant to paying increased taxes on alcohol but they might welcome strategies that prevent the likelihood of their neighbour returning home intoxicated and playing music loudly at night. Harm reduction approaches have gained support from research undertaken in the USA (Midanik et al. 1994), Australia (Stockwell et al. 1996) and Canada (Single and Wortley 1993). This research has demonstrated consistently that drinking problems are more strongly related to the frequency of heavy drinking occasions than to the overall levels of alcohol consumption. Moreover, these studies found that most acute alcohol-related problems, such as driving-related incidents or family and employment difficulties, were experienced by 'moderate' drinkers who drink heavily occasionally. The implication of these studies is that the prevention message is to avoid severe intoxication, or at least problems when you do drink, rather than the more widely used message – 'drink less'.

Interventions aimed at 'high risk' populations

Currently, the most common forms of prevention activity reported are those aimed at groups identified as being at high risk of developing or aggravating alcohol and/or drug problems. Such targets include night-club patrons, prison inmates or young people in local authority care. Through their work in general practice, general and psychiatric hospitals and A&E departments, clinicians have been identified as a key group to do the following:

● effectively identify people at high risk of alcohol and drug problems;
● provide information around alcohol and drugs;

- refer to non-clinical social agencies (e.g. for marital support or debt advice);
- refer people to appropriate substance-use services.

However, lack of training about alcohol and drugs and a perceived lack of time to engage with substance use may limit clinicians' involvement in this area. A survey of medical schools in the UK revealed that on average students received only 14 hours of formal education on substance misuse over their entire 5- or 6-year course (Glass 1989). Research also suggests that doctors see their role as one of dealing with the diseases caused by alcohol misuse rather than dealing with the people who misuse alcohol. In any profession, some practitioners may perceive all people with alcohol and drug problems, pessimistically, as resistant to change. However, evidence indicates otherwise, particularly in relation to alcohol use on which most research has been based. The World Health Organization consistently found, at 6 to 12 months follow-up, that the mean alcohol consumption was approximately 25 per cent less amongst those who had received simple advice about their alcohol use from their doctors as compared with those who just received an alcohol assessment (Richmond and Anderson 1994). Patients who reduce their alcohol intake are also more likely to experience a reduction in their alcohol-related problems. In one study of forty-five UK general practices, Wallace et al. (1988) found a 21 per cent reduction in alcohol consumption amongst those who received advice from their family doctor. Wallace and colleagues estimated that, if the results of this study were extended to all UK general practices, general practitioners could reduce the excessive alcohol consumption of approximately 250,000 men and 67,000 women to moderate levels.

Promising findings have also been reported for alcohol-related interventions undertaken in A&E departments. A 24-month pilot study undertaken in two London A&E departments found that 48 per cent of those patients identified with an alcohol problem attended an appointment the next day to discuss their alcohol misuse. A further development of this study was carried out at St Mary's Hospital in London where a 1-min alcohol test, the Paddington Alcohol Test, was developed and implemented (Smith et al. 1996). Administration of the PAT increased access to counselling by the A&E department's alcohol worker dramatically (tenfold).

Community action approach

Prevention efforts aimed at high-risk populations have been criticized on the grounds that they are focused too heavily on the individual, individual choice and individual change. Advocates of a community perspective to prevention argue that an individual's decision about health related behaviours are interrelated with the community in which they live and that community's norms and customs (Harrison 1996). They maintain that it is very difficult for individuals to sustain change if they are out of step with behaviour in the community in which they live or if their environment is not supportive of that change. Furthermore, in some communities, political factors may make individual change unrealistic. In the community action model, prevention should be aimed at promoting collective changes at a community level. Effective tobacco smoking campaigns have often focused on community as well as individual change (NHS CRD 1999). While tobacco smoking is still a cause of considerable concern, particularly amongst teenagers, there has been a significant reduction in the prevalence of adult smoking. Smoking amongst adult males has declined by 20 per cent over the last 30 years. Interventions have aimed to change community norms and practices through the use of media publicity, health information and policy. The success of these strategies is reflected in surveys showing that, although in 1964 only 43 per cent of the public recognized that smoking was linked to lung cancer, by the late 1980s 80 per cent acknowledged the link between smoking and cancer. Interestingly, during much of this period the price of cigarettes fell in real terms although cigarette consumption did not increase, largely, it is believed because of the impact of community changes to smoking.

Not all public awareness campaigns have been equally successful. A public awareness AIDS prevention campaign which sent information leaflets to all households in the UK resulted in an increase in people's *knowledge* of AIDS, but people's *attitudes* to AIDS sufferers deteriorated and no other behavioural changes were achieved. This was attributed to the campaign's use of fear and shock tactics which can be counterproductive, especially if they are used with young people. A clear example of this was the UK's 1980s media campaign 'Heroin Screws You Up' which featured posters of allegedly unattractive heroin addicts in an effort to deter people. However, the posters attracted cult status and, if anything, glamorized heroin use. Publicity campaigns can be used effectively if:

- they are integrated with other strategies;
- drug and/or alcohol use is not glamorized;
- there is some kind of interaction between those promoting the message and those receiving it;
- the message delivered is clear;
- this message is universally endorsed by those with influence and authority.

Many US campaigns to reduce illicit drug use have not justified the massive efforts involved, such as Substance Abuse a National Emergency (SANE) or Drug Abuse Resistance Education (DARE). However, since the 1980s, it has been clear that certain specific elements of community action can reduce drug use. Nancy Tobler's (1986) analysis of 143 campaigns found only the programmes 'combining positive peer influence with specific skill training' (reappraised by the Campbell Collaboration 2000) had reduced drug using behaviour at all. One of the first programmes to demonstrate reduced use of cannabis, alcohol and cigarettes by young people a year after a community intervention was 'I-Star' in Kansas City (Pentz *et al.* 1989). This multi-component prevention package used schoolteachers, parents, mass media, community leaders, policy makers and above all young people themselves, within an active, social learning model. The package is now better known as the 'Midwestern Prevention Project', and its benefits have been replicated over time and across different urban settings (Pentz 2001). In general, to prevent the onset of substance use, such 'comprehensive and social influence' programmes are the most successful (William Hansen's [1992] review reappraised by the Campbell Collaboration 2000). In the UK, the *Drugs Prevention Advisory Service* is responsible for building up expertise for effective, community-based programmes.

From reading this section, it will have become clear to you that many preventative schemes, in an effort to maximize their intervention's effectiveness, include aspects of different approaches. Indeed, the preventative strategies widely accepted as having the most impact:

* the tobacco smoking campaign,
* the prevention of HIV among drug users in the UK,
* the drink-drive campaign

have included several different components; for example the drink-driving campaign focused on changing the community's attitudes to drink-driving as well as toughening laws on drinking and driving. Other factors which may explain the relative success of these approaches are that they:

* feature a clear, unambiguous message – don't smoke, don't drink or drive, don't share needles or have unsafe sex – unlike the mixed messages given on alcohol use;
* have won considerable support and have influenced society's norms on the behaviour targeted;
* have been endorsed by a wide variety of professional and community groups and have received political support and funding.

Drug action teams

You may feel that each of these approaches has something to offer the A&E department problem presented to you, but that the medical profession has neither the time nor the resources to initiate any approaches beyond the identification and referral of people with alcohol and/or drug problems.

You may (if you are from the UK) see the advantage of involving your local drug action team, providing they are a 'DAT' which addresses alcohol as well as drug problems.

Drug action teams – DATs

- One hundred and five DATs have been set up in the United Kingdom. Each DAT has representatives from seven bodies which typically have different roles and principles (e.g. police reinforcement versus health authority treatment targets and different views on the 'drug problem' and how it should be approached).
- The DATs have a strategic planning and co-ordinating role and are intended to represent their local community on substance misuse problems and the prevention and treatment of these.
- DATs refer to 'drug reference groups', which consist of experts on drugs and alcohol, where they have been established.

The success of DATs is very much dependent on how they work together as multi-disciplinary teams, given that the representatives often have different agendas and do not always perceive each other's role positively; for example, you can imagine that while some DATs' members might support initiatives such as syringe exchange schemes others may be opposed to them. The clear advantage of DATs is that their budgetary powers have given them the ability to implement plans of action and have prevented them from becoming merely a think tank.

In 2001, the UK launched its National Treatment Agency for Substance Misuse. It will be crucial to strike a balance between the NTA's *national* standards for effective, integrated services and the community innovation and responsiveness offered by *local* DATs.

Clinical implications

- Doctors, despite their present lack of training and their pressurized role, can in many cases make a significant difference if they address people's alcohol and drug use with them.
- Studies show effective brief interventions can be introduced in hospital situations. In the long term, prevention approaches are likely to be more effective if they gain the support of the community and the various groups within it.

Case study

In the following case studies, you are presented with some alcohol and drug concerns related to prevention issues that you are likely to encounter with clients.

Patients: Mr and Mrs Brown

We are always reading in the papers about young people and drug use and we get really worried for our three children, two of them are already teenagers and one is not far behind. Is there anything that can be done to prevent them developing alcohol and drug problems? We think they have an alcohol and drug prevention programme at their school.

Your patients have presented you with a difficult question given that adolescent substance use is complex and our knowledge of this area in many ways limited. In many societies, drug use amongst young people is widespread and regarded as normal, making it difficult to make assumptions about such a large sector of the population; for example, some nationwide surveys of young people and drug use in the UK have found that about half of the young people surveyed have tried some drug at least once (Plant and Plant 1992; Judd and Fitch 1998).

However, not all young people who experiment with drugs go on to develop drug-related problems. Researchers have endeavoured to identify what factors place young people at greatest risk of developing drug problems. The key factors identified by the Good Practice Unit for Young People and Drug Misuse (1997) are:

- detachment from family;
- detachment from school;
- detachment from society – its rules of conduct and convention;
- attachment to drug-using peers.

Family influences are important, such as conflict within the family, parents' own use of drugs or tolerance of drugs and emotional, sexual and/or physical abuse. Lack of appropriate discipline or impoverished family circumstances can increase a young person's susceptibility to early drug use, with peer group influences featuring more strongly as the adolescent becomes older. Peer influences are often related to detachment from school. The young person who feels disenchanted with their experience of school is more likely to play truant and in turn to spend time with other truants who are engaged in 'deviant behaviour' including drug use.

According to the Problem Behaviour Theory (Jessor and Jessor 1980), young people who have few attachments to the conventions of society are more likely to use drugs. Richard and Shirley Jessor found in a longitudinal study of 13- to 30-year-olds that the adolescents most likely to engage in drug taking were those: who perceived their parents as unsupportive; whose friends had the predominant influence on their behaviour, and who were focused on personal autonomy (with a low tolerance for society's conventions and a higher tolerance for law breaking). The original studies from the 1970s have been replicated with young people in the 1990s for cannabis use (Donovan 1996) and problem drinking (Donovan *et al.* 1999).

Of course, not all young people exposed to these influences develop drug problems and researchers have sought to identify factors which *protect* young people from drug use. Not surprisingly, these relate closely to the risk factors discussed earlier and can be summarized as follows:

- attachment to at least one parent;
- attachment to school;
- attachment to non-using drug peers;
- onset of drug use delayed.

Considerable research has clearly demonstrated the positive effects of mutual attachment with at least one parent. Likewise, there is evidence to show that appropriate discipline and caring family structures can place young people at less risk of drug use.

More recently, studies have highlighted the importance of young people's attachment to school. Young people who feel a sense of educational competency, who feel connected with their place of education and in particular with a caring supportive teacher, are less likely to use drugs. Furthermore, young people focused on schooling generally have less opportunities to form friendships with drug-using peers, given that these young people are often involved in a range of illicit behaviours and are often not regular school attendees.

The Health Advisory Service (1996) has reviewed this range of factors known to influence substance use during childhood. Together these factors may delay the age at which young people experiment with drugs increasing the chance that it will not begin at all or that it will never lead to problematic use.

A summary of possible interventions

A number of prevention schemes have been established in an effort to prevent drug and alcohol misuse amongst young people. These are mostly school based and typically include drug and alcohol information along with affective components such as decision making. School-based programmes are more likely to be successful if they are integrated into the school curriculum, are facilitated by people the students have an ongoing relationship with (such as teachers) and avoid scare tactics or strategies which inadvertently glamorize drug use. However, some drug and alcohol education programmes only promote, at best, short-term changes to young people's substance use and may effect no change at all (Plant and Plant 1992).

More recently, particularly in the USA, emphasis has been placed on school-based social skills training programmes which include assertiveness training, self-esteem building, stress management and goal setting among other components. Gorman (1996) conducted major reviews of these programmes and found that they had little impact on adolescents' drug- and alcohol-using *behaviour*. Gorman suggests this is because they target only a narrow repertoire of behaviours whilst alcohol and drug use is complex and includes sociocultural factors as well as interpersonal ones. Programmes also risk normalizing drug use, creating the impression that drug use is more widespread than it actually is by overemphasizing the pressures on young people to use drugs.

Peer-led programmes (i.e. programmes where young people take a key role in communicating information about alcohol and drugs to their peers) have shown more positive results. Tobler (1986: see 'Community action approach', p. 283) found in her meta-analysis of adolescent drug prevention programmes that young people only made positive changes to their alcohol and drug *use* in programmes involving their peers.

Several authors have suggested that prevention programmes need to consider factors beyond the young person, peer pressure and decision making, and take into account the realities of family and societal life. Certainly, those programmes which have included parent training, for example, have shown some positive results but again the complexity of adolescent drug and alcohol use is demonstrated with one study showing that, whilst parent training along with involvement in a 6-year school-based prevention programme has an effect on lowering the girls' initiation into drug use, it had no such impact for the boys in the study (O'Donnell *et al.* 1995).

Implications for child development

In your advice to Mr and Mrs Brown, you might emphasize that:

- adolescent drug and alcohol use is complex and our knowledge of what factors prevent adolescents from developing alcohol and drug problems is limited;
- that maintaining close relationships with your children and openly communicating is likely to be helpful;

- encouraging their child's involvement in school related activities may be helpful.

> *Patient: Mrs Smith*
>
> Just before I go, there is one thing. My husband and I usually drink wine with our meals. Do you think this will help our teenage daughter to use alcohol sensibly in the future?

Your patient has presented you with yet another challenging question which is made more difficult by the fact that most research on the relationships between parents' drinking patterns and their children's future use is on the parents' drinking. From this research we know that:

- children of alcoholics are approximately five times more likely to develop alcohol-related problems than their peers;
- the extent to which transmission is genetic, environmental or a combination of both varies according to the type of alcohol problem studied;
- the stress often present in family environments where one or both parents drink could contribute to children developing future alcohol problems.

The very little we know about the influence of parents' (without identified alcohol problems) drinking on their children's future drinking behaviour is that young people do tend to imitate parents' drinking behaviour later.

Implications for parents

In your advice to Mrs Smith, you might emphasize that:

- adolescent drinking and the development of adult drinking patterns are complex and unlikely to be determined by just one factor such as family environment;
- modelling the sensible use of alcohol is more likely to be helpful than unhelpful;
- not all people exposed to parents with alcohol problems develop alcohol problems, again highlighting the complexity of this issue and that the relationship between parental drinking and young people's future drinking patterns cannot be easily predicted.

Working with vulnerable groups

This is becoming one of the most exciting and creative areas of preventive work (e.g. working in outreach to homeless young people or engaging with the teenage clients of a Youth Offending Team). In deprived neighbour-hoods with little social capital, community educational initiatives like 'Sure Start' are beginning to connect with socially isolated, unsupported, teenage, single parents whose young families have been at high risk of alcohol-related harm (Caan 2000). A common lesson in this area of preven-tion is that the interventions need to be intensive and sustained over time, based on a good working knowledge of local conditions and on building partnerships with a variety of 'local players' (Thompson 1999); for example, children in local authority care are at especially high risk of substance misuse, but interactive methods with the children that 'focus on life skills, problem-solving and decision-making' can help (Richardson and Joughin 2000).

The way ahead

The science *Foresight* programme has identified early interventions to prevent drug-related harm as a priority for UK research in education, health and social policy over the next 20 years (Healthcare Panel 2000). Many of the possible community actions to be tested are similar to the developmental, social and environmental interventions under consideration more widely to deliver Mental Health Promotion (NHS Executive 2001). We stand on the brink of integrating many different areas of knowledge – about learning, parenting, school health, workplace opportunities, the physical environment, the psychosocial environmental, citizenship and participation.

Acknowledgements

Additional material for this chapter was provided by Jackie de Belleroche and Woody Caan.

References

Babor, T. F. and Grant, M. (1992) 'WHO collaborating investigators project on identification and management of alcohol related problems. Report on Phase II: A randomised clinical trial of brief interventions in primary care', Copenhagen: WHO.

Birckmayer, J. and Hemenway, D. (1999) 'Minimum-age drinking laws and youth suicide, 1970–1990', *Amer. J. Public Health*, **89**, 365–8.

Caan, W. (2000) 'Sure start in practice', *UpStart!*, **2**, 4.

Campbell Collaboration (2000) 'Crime, drugs and alcohol', *Evidence from Systematic Reviews of Research Relevant to Implementing the 'Wider Public Health' Agenda.* York: NHS CRD.

Donovan, J. E. (1996) 'Problem-behavior theory and the explanation of adolescent marijuana use', *J. Drug Issues*, **26**, 379–404.

Donovan, J. E., Jessor, R. and Costa, F. M. (1999) 'Adolescent problem drinking: Stability of psychosocial and behavioral correlates across a generation', *J. Studies on Alcohol*, **60**, 352–61.

Glass, I. (1989) 'Undergraduate training in substance abuse in the United Kingdom', *Brit. J. Addict.*, **84**, 197–202.

Good Practice Unit for Young People and Drug Misuse (1997) *Drug-related Early Intervention: Developing Services for Young People and Families.* London: SCODA.

Gorman, P. (1996) 'Do school-based social skills training programs prevent alcohol use among young people?', *Addict. Res.*, **4**(2), 190–210.

Hansen, W. B. (1992) 'School-based substance abuse prevention: A review of the state of the art in curriculum, 1980–1990', *Health Educ. Res.*, **7**, 403–30.

Harrison, L. (ed.) (1996) *Alcohol Problems in the Community.* London: Routledge.

Health Advisory Service (1996) *The Substance of Young Needs. Commissioning and Providing Services for Children and Young People Who Use and Misuse Substances.* London: HAS.

Healthcare Panel (2000) *Foresight. Health Care 2020.* London: DTI.

Holt, S., Stewart, I. C., Dixon, J. M., Elton, R. A., Taylor, T. V. and Little, K. (1980) 'Alcohol and the emergency service patient', *Brit. Med. J.*, **281**, 638–40.

Jessor, R. and Jessor, S. (1980) 'A social-psychological framework for studying drug use', *NIDA Research Monograph*, **30**, 102–9.

Judd, A. and Fitch, C. (1998) 'National surveys of drug use', *Drug Use in London.* London: The Centre for Research on Drugs and Health Behaviour.

Midanik, L., Tam, T., Greenfield, T. and Caetano, R. (1994) *Risk Functions for Alcohol-related Problems in a 1998 US National Sample.* Berkeley, CA: California Pacific Medical Center Research Institute, Alcohol Research Group.

NHS CRD (1999) 'Preventing the uptake of smoking in young people', *Effective Health Care*, **5**, Bulletin 5.

NHS Executive (2001) *Making it Happen. A Guide to Delivering Mental Health Promotion.* Leeds: NHS E.

O'Donnell, J., Hawkins, D., Catalona, R., Abbott, R. and Day, E. (1995) 'Preventing school failures, drug use and delinquency among low-income children: long term intervention in elementary schools', *Amer. J. Orthopsychiat.*, **65**, 87–100.

Pentz, M. A. (2001) 'Preventing drug abuse through the community: Multicomponent programs make the difference', paper given at the National Conference on Drug Abuse Prevention Research, NIDA website updated 22 January (the original 1996 conference papers are also available from the National Clearinghouse for Alcohol and Drug Information).

Pentz, M. A., Dwyer, J. H., MacKinnon, D. P., Flay, B. R., Hansen, W. B., Wang, E. Y. and Johnson, C. A. (1989) 'A multicommunity trial for primary prevention of adolescent drug abuse. Effects on drug use prevalence', *JAMA*, **261**, 3259–66.

Plant, M. and Plant, M. (1992) *Risk Takers: Alcohol, Drugs, Sex and Youth*. London: Routledge.

Richardson, J. and Joughin, C. (2000) 'She got involved in prostitution to help pay for her drug habit', *The Mental Health Needs of Looked After Children*. London: Royal College of Psychiatrists.

Richmond, R. L. and Anderson, P. (1994) 'Research in general practice for smokers and excessive drinkers in Australia and the UK. I. Interpretation of results', *Addiction*, **89**, 35–40.

Single, E. and Wortley, S. (1993) 'Drinking in various settings: findings from a national survey in Canada', *J. Studies on Alcohol*, **54**, 591–9.

Smith, S. G., Touquet, R., Wright, S. and Das Gupta, N. (1996) 'Detection of alcohol misusing patients in accident and emergency departments: The Paddington alcohol test (PAT), *J. Accident and Emergency Med.*, **13**, 308–12.

Stockwell, T., Hawks, D., Lange, E. and Rydon, P. (1996) 'Unravelling the preventive paradox for acute alcohol problems', *Drug and Alcohol Rev.*, **15**, 7–15.

Thompson, A. (1999) 'Smoking out the problem', *Community Care*, 22 April, 20–1.

Tobler, N. (1986) 'Meta analysis of 143 adolescent drug prevention programs: Quantitative outcome results of program participants compared to a control or comparison group', *J. Drug Issues*, **16**, 537–67.

Wallace, P., Cutler, S. and Haines, A. (1988) 'Randomised control trial of general practitioner intervention in patients with excessive alcohol consumption', *Brit. Med. J.*, **287**, 663–8.

Walsh, M. E. and Macleod, D. (1983) 'Breath alcohol analysis in accident and emergency department', *Injury*, **15**, 62–6.

Horizon scanning

In this book, we have tried to provide an introduction to the neuroscience and psychology of addiction, as we think that it is important to chart the convergence between biological knowledge, on the one hand, and the needs of individuals and society, on the other in order to understand the problems of addiction. We have also tried to indicate that the problems associated with substance use can respond to a range of appropriate interventions.

Looking ahead

There is no question that the illegal drugs trade is booming. In the UK, expenditure on illegal drugs is equivalent to approximately 2 per cent of total consumer spending (1.1 per cent GNP in 1996) and a similar amount is spent on legitimate drugs, such as the wine and spirits trade (D. Butler [1998] *Accountancy*, November, 34–7). The undesirable consequences of substance misuse are inevitably increasing in the wake of this expenditure.

Economic implications of heroin addiction are illustrated by the patterns of crime: £20,000 per annum is the minimum required to provide 0.5 g per day. Therefore, by putting an addict into effective treatment, there is a massive financial saving for society through reducing the criminal activities needed to support the addict's habit.

The social impact of substance use is huge and the effects of drugs extend from the acute consequences of all drugs (including nicotine) to their extended influences on development, even from utero, through childhood and into adulthood.

There has been a new pharmacological understanding stemming from medicinal drug development but there is potential for the misuse of new preparations such as cognitive enhancers which have powerful effects on attention. Even herbal preparations whose sales are rocketing in health shops and supermarkets are potential sources of risk that may create problems of toxicity and dependence on a large scale. Some preparations are even marketed as herbal 'Ecstasy'. Hormones like anabolic steroids have already entered the drug culture and others may emerge in the future, like analogues of vasopressin. The extent of associated problems of this type of drug misuse are not known as much is to be found out about their patterns of use, psychological effects and potential for dependence. The same is true for new lifestyle drugs such as viagra.

Scanning the horizon, it is very likely that new designer hallucinogens and opiates will appear and, therefore, we need to be prepared for the unknown – new classes of drug, new technologies and commonplace substances administered in different forms and by different routes.

Scientific frontiers

The reasons for an early and rapid onset of problems in some teenagers and the genetic vulnerability of some populations have not been clear. You cannot always understand escalating use in biological terms alone (e.g. drink problems in pub landlords or other occupational groups with an increased risk of dependence). A better public health perspective on risk and resilience should emerge as more research is undertaken using an interdisciplinary approach.

The underlying nature of craving and relapse has been little understood, but as more and more researchers take a long-term view of the development of dependence and the process of recovery, we may come to grasp what it is that makes addiction so compelling. To date, developments have been made in key areas such as:

- psychological management of craving;
- anti-craving drugs;
- cognitive therapy and the promotion of emancipatory change;
- improved access to employment and other life chances;
- social interventions at family and community levels.

Further advances in these areas are likely to be important in the future. These could range from interventions in *early years enrichment* and *neighbourhood renewal* which could prevent drug use in young people to interventions offering adults with severe dependence the chance to *divert* from crime or homelessness to effective rehabilitation and living in a drug-free environment.

The need for improved management of 'dual diagnosis' clients will drive more *integrated* care planning which can respond to overlapping mental health and addiction needs.

Primary prevention, of course, remains the major objective for the future. As with smoking, the best preventative measure is to avoid starting in the first place, because anything else has a relatively marginal effect. The substantiated scientific analysis of the effects of smoking on lung cancer, which drove strong government measures, are now resulting in greatly improved life expectancy and reduced disability. The problems with illegal drugs will similarly benefit from greater public awareness and by providing the right *context* for the individual to make their choices. However, more specialized approaches such as 'Drug Action Teams' are being developed to reach children and 'at risk' groups more directly, with specifically targeted methods.

At the molecular level, we may just see *biotech* companies like Cantab succeed in developing practical vaccinations and interventions for nicotine and cocaine use. At the policy level, we may see strategies such as 'Tackling Drugs to Build a Better Britain' creating new partnerships between agencics and between professionals and patients that will promote co-operation across all the stages and pathways to change.

Jackie de Belleroche and Woody Caan

Index